American Book Company
The Standards Experts

COMMON CORE IN GRADE 3

MATHEMATICS

2016–2017 EDITION

Author
Lindsay Shaughnessy

Curriculum Designer
Bryan Portnoy

AMERICAN BOOK COMPANY
P. O. BOX 2638
WOODSTOCK, GEORGIA 30188-1383
TOLL FREE 1 (888) 264-5877 PHONE (770) 928-2834
TOLL FREE FAX 1 (866) 827-3240
WEB SITE: www.americanbookcompany.com

ACKNOWLEDGEMENTS

The author would like to gratefully acknowledge the technical contributions of Becky Wright. We want to thank Samuel Rodriguez for his editing expertise; Billie Stewart, Tabatha Martin, and Joshua Tompkins for their proofreading expertise; and Lauren Anderson and Ryan Guyer for their expertise in developing many of the graphics for this book.

© 2016 Published by American Book Company
PO Box 2638
Woodstock, GA 30188-1383

Table of Contents

American Book Company

THE STANDARDS EXPERTS

Dear Student

Welcome to your new American Book Company Learning Program!
This book has been created especially for you. Our writers have covered
100% of the standards and concepts as clearly and simply as possible.

As with every ABC Learning Program, this book now comes with an eBook
to expand your educational experience! There are instructions on the following
pages that show you how to access your eBook. We hope that having this digital
copy will allow you to go deeper into your studies.

We look forward to hearing of your success!

eBooks
Access your eBooks now!

E-Mail Address

Password

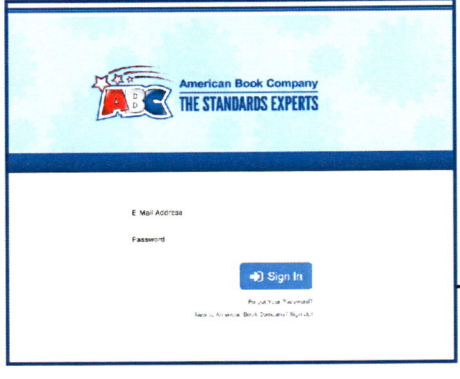

➡] Sign In

Forgot Your Password?

New to American Book Company? Sign Up!

1 - Go to: http://americanbookcompany.com/ebooks

2 - Click on **"New to American Book Company? Sign Up!"**
Then Follow steps in the Registration Process using RedShelf.

3 - After registration you will be taken to your **"My Shelf"** page.
Here you will be able to redeem the access codes for you eBooks via the
"Redemption Code" box. Each book will show up in **"My Shelf."**

Your eBook codes can be located
on the front cover of your workbook.

Use this code to activate your eBook at:

YZH37MVESXC63TU8

http://americanbookcompany.com/ebooks

eBook Features

- Linked table of contents
- Searchable text for keywords
- Wikipedia and Google key term define
- Note-taking, highlighting,
  and free draw tool

Preface

Common Core in Grade 3 Mathematics will help you review and learn concepts and skills related to 3th grade mathematics. **The materials in this book are based on the Common Core Standards in mathematics, coordinated by the National Governors Association Center for Best Practices and the Council of Chief State School Officers. The complete list of standards and answers to exercises are located in the Teacher Guide.**

Common Core in Grade 3 Mathematics includes Depth of Knowledge levels for four content areas based on Norman Webb's Model of interpreting and assigning depth of knowledge levels to both objectives within standards and assessment items for alignment analysis.

The four levels of Depth of Knowledge:

Level 1: Recall and Reproduction (DOK 1)

Questions at this level (DOK 1) include the recall of information such as facts, definitions, or simple procedures, as well as performing a simple algorithm or carrying out a one-step, well-defined, straight forward procedure. A few example DOK level 1 questions are listed below.

- State the associative property of multiplication.
- Measure the perimeter of a figure.
- Calculate 4.3 + 8.5.

Level 2: Skills and Concepts/Basic Reasoning (DOK 2)

DOK level 2 questions involve some mental processing beyond a habitual response. They require students to make some decisions as to how to approach the problem, as well as to classify, organize, estimate, make observations, collect, display, and compare data. A few example DOK 2 level questions are listed below.

- Interpret the bar graph to answer questions about a given population.
- Classify different types of polygons based upon their characteristics.
- Extend an algebraic pattern.

Level 3: Strategic Thinking/Complex Reasoning (DOK 3)

This level (DOK 3) includes problems that require reasoning, planning, and using evidence and higher levels of thinking beyond what was required in DOK levels one and two. This level requires students to explain their thinking, and cognitive demands are more complex and abstract. DOK 3 demands that students use reasoning skills to draw conclusions from observations and make conjectures. Some examples of level 3 DOK questions are listed below.

- Explain how you can determine if two triangles are similar.
- Formulate an expression to determine the next few terms in a pattern.
- Construct and conduct a survey and analyze the results to determine the most popular movie genre.

Level 4: Extended Thinking/Reasoning (DOK 4)

DOK level 4 questions include things such as complex reasoning, planning, and developing. Student thinking will most likely take place over an extended period of time and will include taking into consideration a number of variables. Students should be required to make several connections and relate ideas within the content area or among other content areas by selecting one approach among many alternatives on how a situation should be solved. At this level students will be expected to design and conduct their own experiments, make connections between findings, and relate them to concepts and phenomena together. A few example problems are presented below:

- Explore real world phenomena of Cartesian plans and create a report to present your findings.

- Connect your knowledge of integers to the plate tectonics of Earth.

- Analyze common game pieces (i.e. dice, spinners, etc.) to determine their fairness based upon what you know about probability of events by designing and carrying out your own experiment.

About the Author:

Lindsay Shaughnessy graduated cum laude from Kennesaw State University with a bachelor's of science in mathematics. Throughout college, she worked on campus as a peer mentor and supplementary instructor with Early Start Bridge Academy, a bridge program for first-year university students. Lindsay also tutored nontraditional students at the Lifelong Learning Center and all students in the Math Lab. She was awarded the National Tutor of the Year award in 2013. She plans to attend graduate school in order to teach at the university level.

About the Curriculum Designer:

Bryan Portnoy has spent over 15 years in education, most of them teaching mathematics at the high school level. He has a Bachelor of Arts degree in mathematics from SUNY Albany and a master's in educational leadership from Georgia State University. Bryan's focus in the classroom was helping all students achieve by meeting the needs of each individual student through his or her own learning style. Bryan is currently serving as the math curriculum director.

Chart of Standards

The following chart correlates the questions in each part of the pretest and the post test to the Common Core State Standards. These test questions are also correlated with chapters in *Common Core in Grade 3 Mathematics*. **Note:** Some question numbers appear under multiple standards if the questions cover more than one claim or assessment target.

Common Core Standard	Chapter	Pretest	Post Test
3.OA.1	3	16	17
3.OA.2	4	6	6
3.OA.3	3,4	1, 20, 40	1, 20
3.OA.4	3,4	18	19
3.OA.5	3,4	26	8, 26
3.OA.6	4	13	14
3.OA.7	3,4	8, 25	11, 27, 34
3.OA.8	3,4	12, 22, 39	13, 23, 39
3.OA.9	2	10	12, 38
3.NBT.1	1,2	9	9
3.NBT.2	2	15, 29, 36, 37	16, 23, 30
3.NBT.3	3	–	–
3.NF.1	5	14	15, 29
3.NF.2	5	–	–
3.NF.3	5	–	–
3.MD.1	6	33	34, 40
3.MD.2	6	27	28
3.MD.3	7	4, 19, 31, 35, 38	4, 21, 32
3.MD.4	5,6,7	30	31
3.MD.5	8	–	35
3.MD.6	8	16	17
3.MD.7	8	2, 23	2, 24
3.MD.8	8	3, 5, 24, 34	3, 5, 25, 35
3.G.1	8	7, 11, 17, 32	7, 10, 18, 33, 37
3.G.2	8	21, 28	22, 36

Depth of Knowledge Chart

The following chart shows the correlation of all test questions to Depth of Knowledge levels. The abbreviation DOK is used to denote the level on each test question.

	DOK 1	DOK 2	DOK 3	DOK 4
Pretest	1, 2, 5, 9, 13, 18, 22, 25, 27, 28	3, 4, 7, 10, 11, 14, 15, 16, 17, 20, 21, 23, 26, 30, 31, 32, 35, 36, 40	6, 8, 12, 19, 29, 33, 34, 37, 38, 39	24
Post Test	1, 2, 5, 9, 14, 15, 19, 23, 27, 28, 29, 37	3, 4, 7, 8, 10, 12, 16, 18, 20, 22, 24, 26, 31, 32, 33, 38, 39, 40	6, 11, 13, 17, 21, 30, 34, 35, 36	25

Pretest
Part 1

Follow the directions for each question.

1   Circle true or false for each question.

    A $3 \times 5 = 60 \div 4$ True   False

    B $4 \times 6 = 50 \div 2$ True   False

    C $5 \times 7 = 80 \div 2$ True   False

    D $8 \times 4 = 64 \div 2$ True   False

3.OA.3 DOK 1

2   The rectangle below is divided into square units. What is the area of the rectangle?

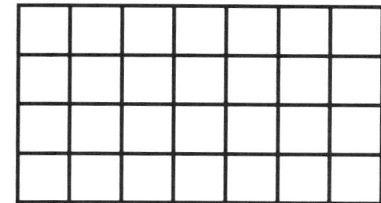

3.MD.7A DOK 1

3   Jim wants to find the perimeter of the hexagon below. Show how Jim should set up his problem and the correct answer.

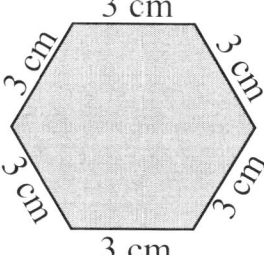

3 cm
3 cm
3 cm
3 cm
3 cm
3 cm

3.MD.8 DOK 2

4   Look at the bar graph below. How many more woolley-bear caterpillars did John catch than Amy? Choose the correct answer.

Woolley-Bear Caterpillars Caught

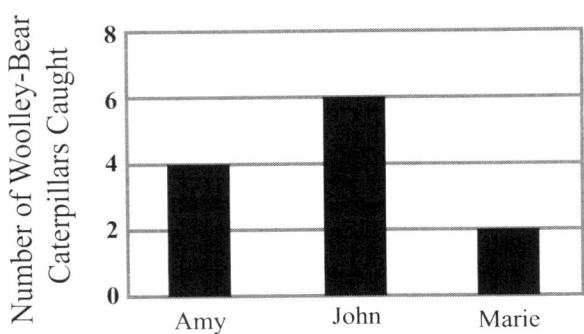

A   6

B   4

C   3

D   2

3.MD.3 DOK 2

5   Anna wants to find out how long her bedroom is. What tool should Anna use to measure the length of her bedroom? Write your answer on the line below.

_____

3.MD.8 DOK 1

6   Alan has $3.00 in his piggy bank. He needs $6.00 to buy a toy he wants. He can earn $1.00 for washing 4 windows for his mother. How many windows will Alan need to wash to have enough money to buy the toy? Write your answer on the line below.

_____

3.OA.2 DOK 3

7   Write the names of the shapes on the lines below the pictures. Then write how many angles each shape has.

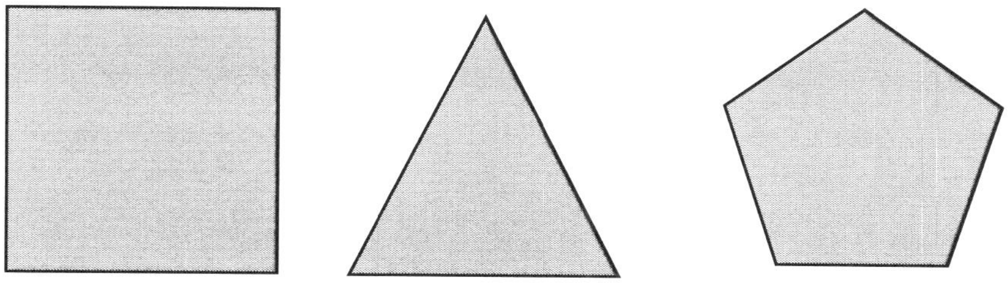

Name: _____   _____   _____

Number of Angles: _____   _____   _____

3.G.1 DOK 2

8   The chart below shows the ages of the Smith family.

Name	Age
Father	40
Mother	35
Alice	10
Jack	7

Answer each of the four questions below. Write your answer on the lines.

A Who is five times older than Jack?   _____

B Who is four times older than Alice?   _____

C What are the ages of the father and mother together?   _____

D What is the age of the father less the age of Alice?   _____

3.OA.7 DOK 3

9 Round the number 6,374 to the nearest hundred. Write your answer on the line below.

_____

3.NBT.1 DOK 1

10 All of the number patterns below follow the same rule. Fill in the missing numbers for each pattern.

A 3, 6, 9, 12, _____

B 18, 21, 24, _____, 30, 33

C 36, 39, 42, _____, 48, 51

D 54, _____, 60, 63, 66

E 69, 72, _____, 78, 81

3.OA.9 DOK 2

11 Which shape has more sides than a square? Choose the correct answer.

A triangle

B pentagon

C rectangle

D trapezoid

3.G.1 DOK 2

12 Amy is rearranging the books on her shelves. She has 7 books on one shelf, 11 books on another shelf, and 0 books on a third shelf. She wants to put an equal number of books on the 3 shelves. How many books should she put on each shelf? Show your work and answer on the line below.

_____

3.OA.8 DOK 3

13 $99 \div N = 11$ and $N \times 6 = 54$. What is the value of N?

_____

3.OA.6 DOK 1

14 Bob has 24 pennies in his piggy bank. He gives 18 of the pennies to his little brother. What fraction of the 24 pennies does he have left? Circle the correct answer.

$$\frac{18}{24} \qquad \frac{3}{24} \qquad \frac{8}{24} \qquad \frac{6}{24}$$

3.NF.1 DOK 2

15 Jill was asked to add 4,231 + 1,705. She lined up the two numbers over each other like this:

$$\begin{array}{r} 4{,}231 \\ + \ 1{,}705 \\ \hline \end{array}$$

Which column should Jill add first? Tom said she should add the thousands column first because it is closest to the plus sign. Sara said she should start by adding the ones column first. Who is correct? Write the name on the line below.

_____

3.NBT.2 DOK 2

16 Look at the square divided into smaller squares below. What information is assumed about the picture? Assumed means something you can count on as true. Circle all of the correct answers.

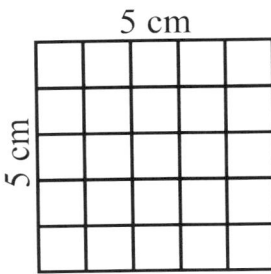

5 cm

5 cm

A All of the squares have equal length sides.

B There are 36 smaller squares inside the larger square.

C The area is 25 square cm.

D The perimeter can be figured by adding the length of the 4 sides:

5 + 5 + 5 + 5 = 20.

3.MD.6 and 3.OA.1 DOK 2

17 Compare the two shapes below.

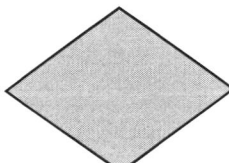 

Choose the sentence below that correctly compares the two shapes.

A The rhombus has fewer sides than the rectangle.

B The rhombus and the rectangle have the same number of sides and angles.

C The rectangle has fewer angles than the rhombus.

D The rectangle has more sides than the rhombus.

3.G.1 DOK 2

18 Circle **all** of the fact families for the numbers 3, 7, and 21.

   A $3 \times 7 = 21$

   B $21 + 7 = 28$

   C $21 \div 3 = 7$

   D $7 \times 3 = 21$

   E $21 - 3 = 18$

   F $21 \div 7 = 3$

3.OA.4 DOK 1

19 The bar graph below shows the favorite flavor of jelly beans of one 3rd grade class. Answer the questions below the graph.

Favorite Jelly Beans of 3rd Graders

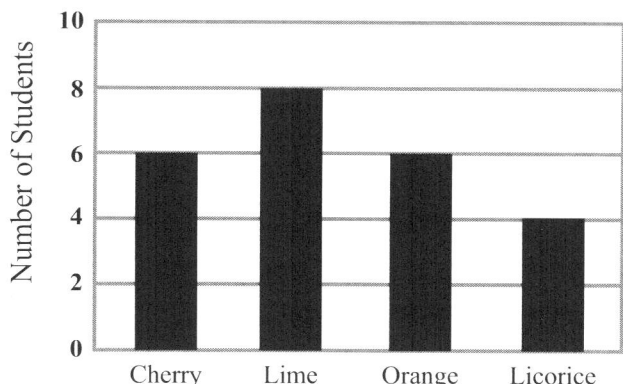

Write your answers on the lines.

A How many students like cherry jelly beans best? _____

B How many more students like lime jelly beans than licorice jelly beans? _____

C How many students like either orange or licorice jelly beans? _____

D Which two flavors were chosen by 6 students each? _____

E How many students like either cherry or lime jelly beans? _____

3.MD.3 DOK 3

20 There are 16 ounces in every pound. How many ounces are there in a 3-pound package of chicken?

_____

3.OA.3 DOK 2

21 Mrs. Smith drew a picture of the tiles on her kitchen floor. She bought a rug that will cover 6 of the tiles. What fraction of her floor can she cover with the rug? Choose the correct answer.

$\dfrac{10}{30}$ $\dfrac{6}{30}$ $\dfrac{20}{30}$ $\dfrac{15}{30}$ $\dfrac{24}{30}$

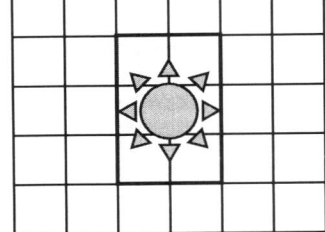

3.G.2 DOK 2

22 What is N in the problem $N - 8{,}067 = 1{,}510$? Choose the correct answer.

A $N = 9{,}567$

B $N = 6{,}557$

C $N = 9{,}077$

D $N = 9{,}577$

3.OA.8 DOK 1

23 Madison was asked to find the area of the square of flowered paper shown below. Show how Madison should set up and solve her problem. Write your answer on the line below.

9 inches

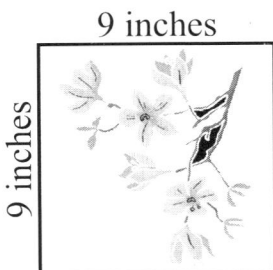

9 inches

3.MD.7b DOK 2

Page 8

24 Andrew and his father built a rectangular sandbox that measures 6 feet by 7 feet for his little brother. His little brother is older now and needs a larger sandbox. Andrew's father says the new sandbox will have a perimeter 12 feet more than the old sandbox. They reuse the wood from the 7-foot sides of the sandbox and buy new wood for the other sides. How long will the other sides of the rectangular sandbox be? Write your answer on the line below.

_____

3.MD.8 DOK 4

25 Circle all the problems below that are solved correctly.

A $56 \div 8 = 8$

B $35 \div 7 = 5$

C $72 \div 8 = 9$

D $90 \div 9 = 11$

E $49 \div 7 = 7$

3.OA.7 DOK 1

26 Olivia wants to know how many eggs are in 6 cartons that are all the same size. Her friend, Abby, says she should count how many eggs are in one carton, and add 6 to that number. Her brother, Joe, says she should count how many eggs are in one carton and multiply by the number of cartons. Who is correct? Write the name on the line below.

_____

3.OA.5 DOK 2

27 Which of the following should be measured in grams, and which should be measured in kilograms? Circle the correct answer for each line.

A button   **grams**   **kilograms**

B dump truck carrying full load   **grams**   **kilograms**

C mass of a woman   **grams**   **kilograms**

D a newborn hamster   **grams**   **kilograms**

3.MD.2 DOK 1

28 Which of the fraction models below correctly shows $\frac{3}{4}$ of the model shaded?

A

B

C

D

E

3.G.2 DOK 1

29 Jacob was trying to solve the problem $(2 + 5) - 3$. First, he added $2 + 5 = 7$. Then, he added 3 so that $7 + 3 = 10$. He did not get the correct answer. Where did he go wrong? Write a sentence explaining where Jacob went wrong, and give the correct answer on the lines below.

_____

_____

3.NBT.2 DOK 3

30 Elizabeth wants to measure her sister's height and weight, and she wants to find out how many more days until her sister's birthday. What tools should Elizabeth use to find these measurements of her little sister? Write your answers on the lines below.

3.MD.4 DOK 2

31 The line graph below shows the number of puppies brought into the Friendly Pet Clinic from January through June.

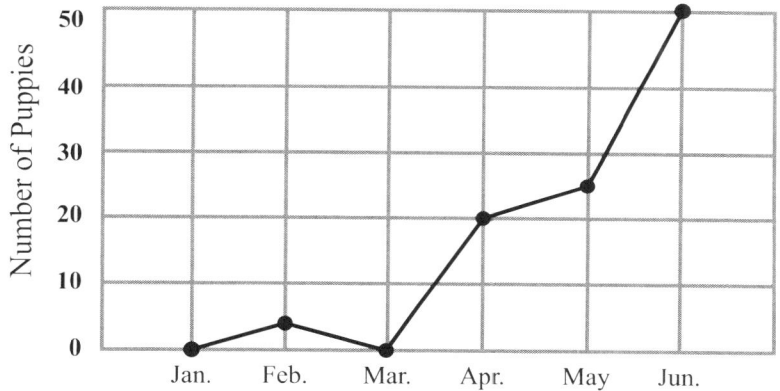

Number of Puppies Brought into the Friendly Pet Clinic From January - June

Write a sentence about the amount of puppies brought into the Friendly Pet Clinic between January and June. Write your answer on the line below.

3.MD.3 DOK 2

32 Suppose your teacher gives you a piece of paper and tells you it is a square. Without using a ruler or a protractor to verify that it is a square, what would you know for sure about the square? Write two things that define a square on the lines below.

3.G.1 DOK 2

33 William wants to earn $48.00 to buy a game system from his cousin. He earns $8.00 raking a lawn. How many lawns will he need to rake in order to buy the game system from his cousin? Set up the problem on the line below and solve.

_____

How many days do you think it will take William to rake that number of lawns and why? Write your answer on the line below.

_____

3.MD.1 DOK 3

34 Find the area of the figure below.

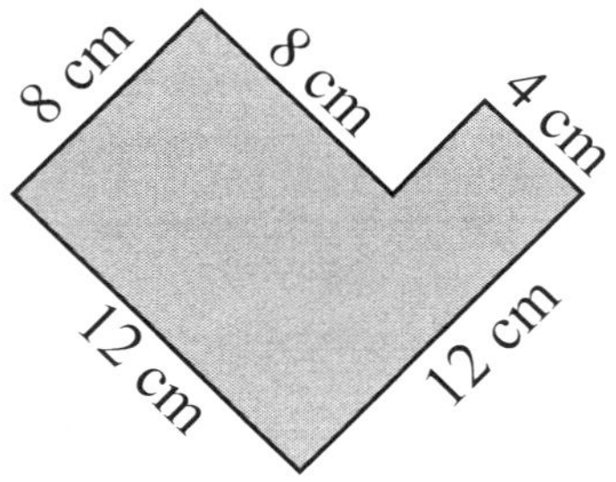

Use the lines below to show your work for each step. Show your answer in square cm.

_____

_____

_____

_____

3.MD.8 DOK 3

Performance Task

Trixie is donating teddy bears to a charity. The bar graph shows the number of teddy bears donated in each of four weeks in October.

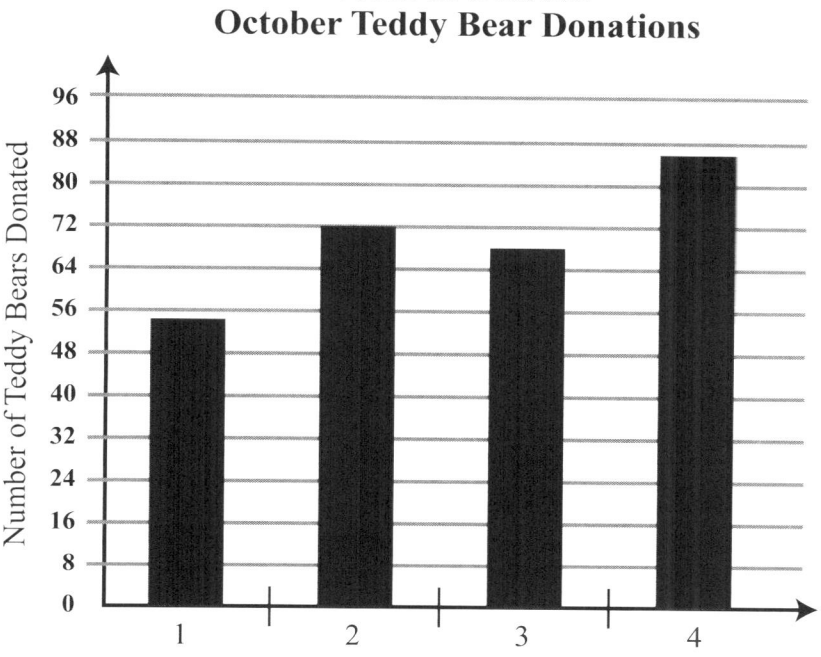

October Teddy Bear Donations

Use the October Teddy Bear Donations bar graph to complete this task.

35 The bar graph shows how many teddy bears Trixie donated in October. Complete the table to show how many teddy bears Trixie donated each week.

October Teddy Bear Donations

Week	Number of Teddy Bears Donated
Week 1	
Week 2	
Week 3	
Week 4	

3.MD.3 DOK 2

36 How many total teddy bears did Trixie donate to charity in October?

3.NBT.2 DOK 2

37 Trixie also donated teddy bears for 4 weeks in November. She compares her weekly donations in October to her weekly donations in November.

For week 1, she donated 14 fewer teddy bears in November than in October.
For week 2, she donated 18 more teddy bears in November than in October.
For week 3, she donated 17 more teddy bears in November than in October.
For week 4, she donated 26 fewer teddy bears in November than in October.

Complete the table to show how many teddy bears Trixie donated each week in November.

November Teddy Bear Donations

Week	Number of Teddy Bears Donated
Week 1	
Week 2	
Week 3	
Week 4	

3.NBT.2 DOK 3

38 Create a picture graph that shows the number of teddy bears Trixie donated each week in November. Use the November Teddy Bear Donations table from number 37 to complete this task.

A. Select the key you will use.

    🧸 = 5 bears   🧸 = 10 bears   🧸 = 20 bears

B. New picture graph

November Teddy Bear Donations	
Week	Number of Teddy Bears Donated
Week 1	 🧸🧸🧸🧸🧸🧸🧸🧸🧸🧸
Week 2	 🧸🧸🧸🧸🧸🧸🧸🧸🧸🧸
Week 3	 🧸🧸🧸🧸🧸🧸🧸🧸🧸🧸
Week 4	 🧸🧸🧸🧸🧸🧸🧸🧸🧸🧸

In the graph, you may use whole and half pictures for each teddy bear.

- First, select the scale in the key that you will use for your graph.
- Then, shade in the teddy bears to create the graph.

3.MD.3 DOK 3

39 Did Trixie's total teddy bear donations increase or decrease in November compared to October? Explain how you determined your answer.

3.OA.8 DOK 3

40   At the end of each week, Trixie helps box up teddy bears to send to the charity. Each box can hold 5 teddy bears.

Part A

Complete the table to show how many boxes Trixie needs each week of November.

November Teddy Bear Donations

Week	Number of Teddy Bears Donated	Number of Boxes
Week 1		
Week 2		
Week 3		
Week 4		

Part B

Trixie has 50 boxes in her garage. Will she have enough boxes to hold the total number of teddy bears donated in November? Explain why or why not.

3.OA.3 DOK 2

Chapter 1
Number Sense

This chapter covers the following Grade 3 standard:

	Content Standard
Number and Operations in Base 10	3.NBT.1

1.1 Place Value (DOK 1, 2, 3)

A **place value** describes the value of a **digit** (any number from 0 to 9) in a number. Take a look at the number 1,863.

Thousands	Hundreds	Tens	Ones
1,000	100	10	1
1 ,	8	6	3

Ones: The 3 is in the ones column.

Tens: The 6 is in the tens column. This means there are 6 sets of 10.

Hundreds: The 8 is in the hundreds column. This means there are 8 sets of 100.

Thousands: The 1 is in the thousands column. This means there is one set of 1,000.

Therefore, $1 \times 1,000 + 8 \times 100 + 6 \times 10 + 3 \times 1 = 1,000 + 800 + 60 + 3 = 1,863$.

Write the place value of 2 for each number. Then, give the value. The first one has been done for you. (DOK 1)

1. 2,457 ___thousands, 2,000___

2. 4,278 _____

3. 952 _____

4. 8,724 _____

5. 6,271 _____

6. 523 _____

7. 1,782 _____

8. 2,933 _____

Using the number 6,943, write which digit is in the given place value. (DOK 1)

9. ones place:_____

10. tens place: _____

11. hundreds place: _____

12. thousands place _____

Using the number 2,441, write which digit is in the given place value. (DOK 1)

13. ones place:_____

14. tens place: _____

15. hundreds place: _____

16. thousands place: _____

1.2 Numbers in Word Form (DOK 1, 2)

Numbers can be written in standard form and word form. **Standard form** is writing a number using digits (283). **Word form** is writing a number using words (two hundred eighty-three). This section will focus on writing numbers in word form.

Example 1: Write 9,068 in word form.

Step 1: The 9 in the thousands place represents 9,000.

Step 2: Skip over the 0s because they do not hold any value. We do not write out the 0s.

Step 3: The 6 in the tens place represents 60.

Step 4: The 8 in the ones place represents 8.

Step 5: Write the number in word form.

Answer: Nine thousand, sixty-eight.

Rewrite the numbers in word form. (DOK 1)

1. 2,778 _____

2. 16 _____

3. 5,591 _____

4. 644 _____

5. 3,701 _____

6. 777 _____

7. 6,217 _____

8. 8,342 _____

Rewrite the numbers in standard form. (DOK 2)

9. two thousand, ten

    _____

10. five hundred twenty-nine

    _____

11. eighty-four

    _____

12. seven thousand twelve

    _____

13. nine hundred forty-one

    _____

14. thirty-three

    _____

15. two hundred sixteen

    _____

16. forty-nine

    _____

17. one hundred fifteen

    _____

18. six thousand, nine hundred twenty-six

    _____

19. three thousand, two hundred fifty-seven

    _____

20. one thousand, three hundred thirty-three

    _____

21. four thousand, eight hundred fourteen

    _____

22. nine thousand, nine hundred ninety-nine

    _____

23. two thousand, four hundred eighteen

    _____

24. three thousand, one hundred nineteen

    _____

25. seven thousand, two hundred thirty-one

    _____

26. five thousand, six hundred ninety-six

    _____

1.3 Numbers in Expanded Form (DOK 2,3)

To write a number in **expanded form**, separate each digit by place value. Place the number in front of its place value. Finally, rewrite the number as an addition problem. For example, the number 1,111 is in standard form. It is made up of the following parts: 1 thousand + 1 hundred + 1 ten + 1 one. Expanded, this is written as 1000 + 100 + 10 + 1.

Example 1: Write the number 4,682 in expanded form.

Thousands	Hundreds	Tens	Ones
4	6	8	2

The expanded form of 4,682 is 4000 + 600 + 80 + 2.

Example 2: The expanded form of 3,112 = 3,000 + 100 + 10 + 2.

Write the following in expanded form. (DOK 2)

1. 37 _____

2. 4,012 _____

3. 9,772 _____

4. 1,644 _____

5. 8,717 _____

6. 594 _____

7. 18 _____

8. 622 _____

Write the following in standard form. (DOK 2)

9. $600 + 90 + 1$ _____

10. $8,000 + 600 + 50 + 4$ _____

11. $400 + 70 + 2$ _____

12. $1,000 + 400 + 30 + 8$ _____

13. $5,000 + 900 + 20 + 4$ _____

14. $9,000 + 200 + 40 + 5$ _____

15. $4,000 + 2$ _____

16. $6,000 + 700 + 7$ _____

17. $1,000 + 100 + 10$ _____

18. $50 + 2$ _____

19. Michele is writing numbers in expanded form. She knows that there are 365 days in a year. How should Michele write 365 in expanded form? _____

(DOK 3)

20. Ernesto read that there are 6,946 kinds of insects in a state park. Ernesto expanded 6,946 to $6,000 + 900 + 60 + 4$. Explain why Ernesto's answer is wrong. Correct his mistake.

_____

1.4 Modeling Numbers (DOK 1,2)

Models and drawings are another way to represent numbers.

Example 1: Use the model to name the number.

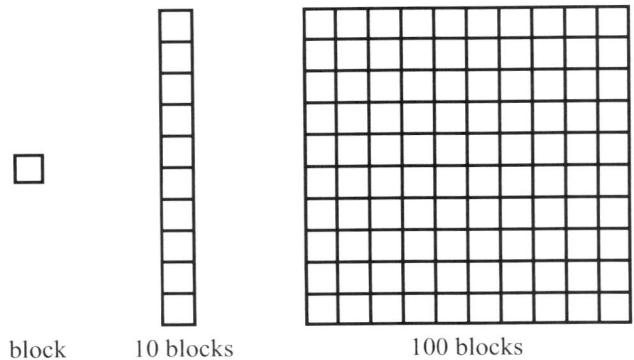

1 block   10 blocks   100 blocks

Step 1: Look at the model. What units are shown?

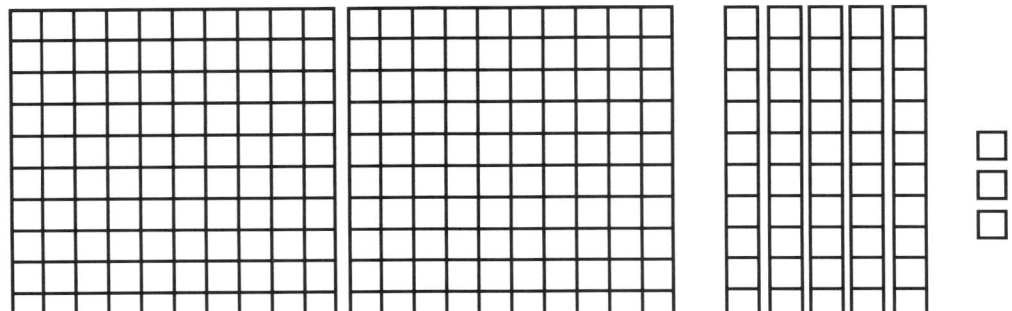

Hundreds, tens, and ones are shown by the blocks above.

Step 2: Start with the largest place value.

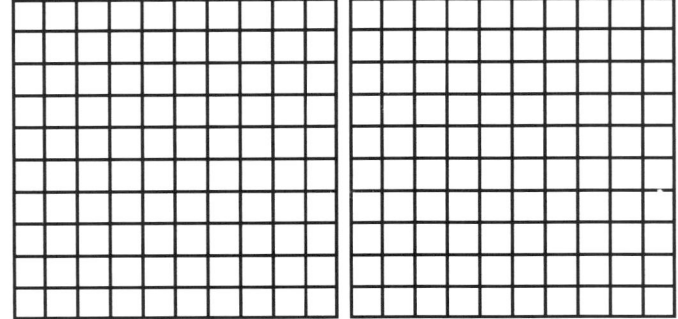

There are two blocks of 100, or 200.

Step 3: Look at the next largest place value shown.

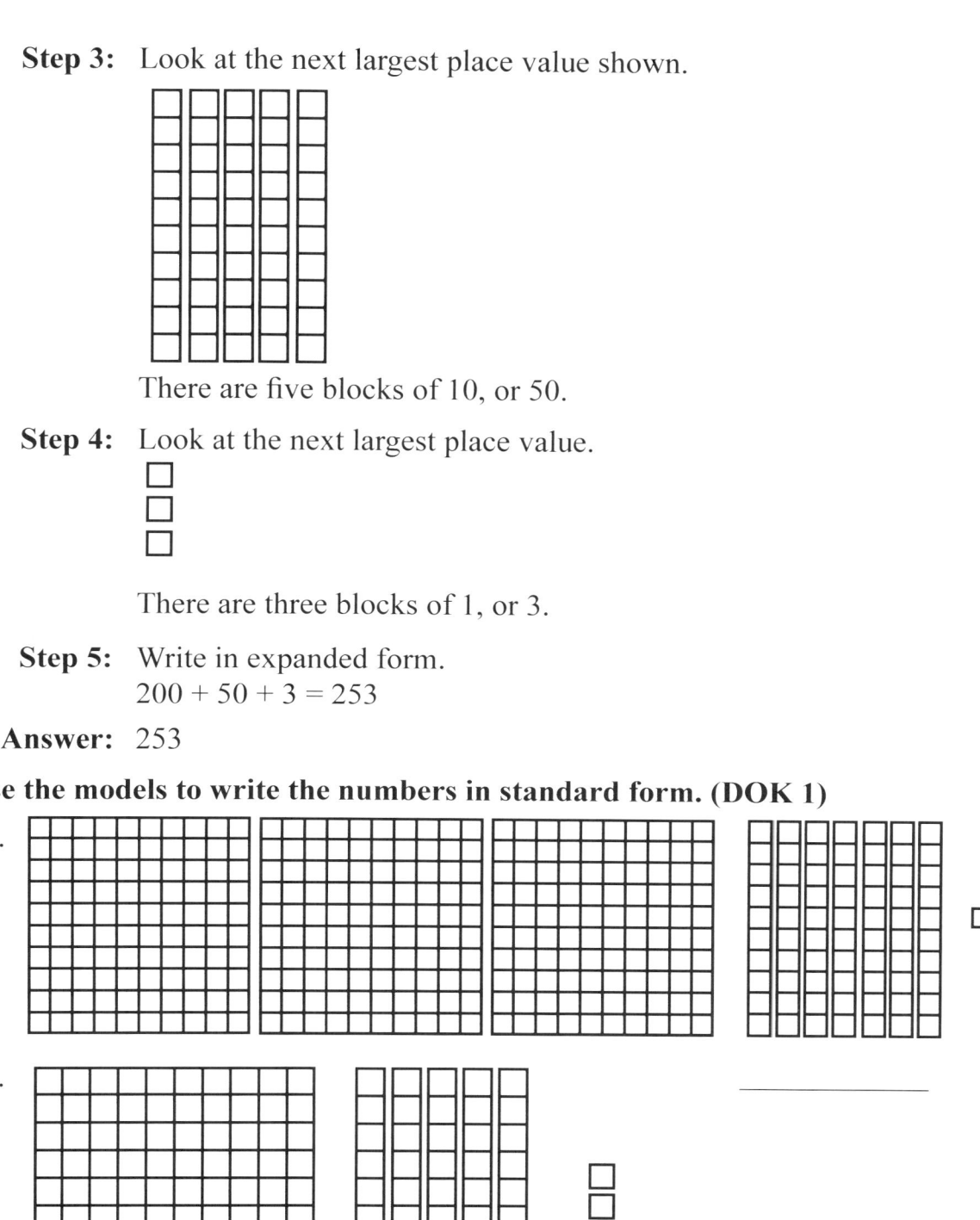

There are five blocks of 10, or 50.

Step 4: Look at the next largest place value.

There are three blocks of 1, or 3.

Step 5: Write in expanded form.
$200 + 50 + 3 = 253$

Answer: 253

Use the models to write the numbers in standard form. (DOK 1)

1.

2.

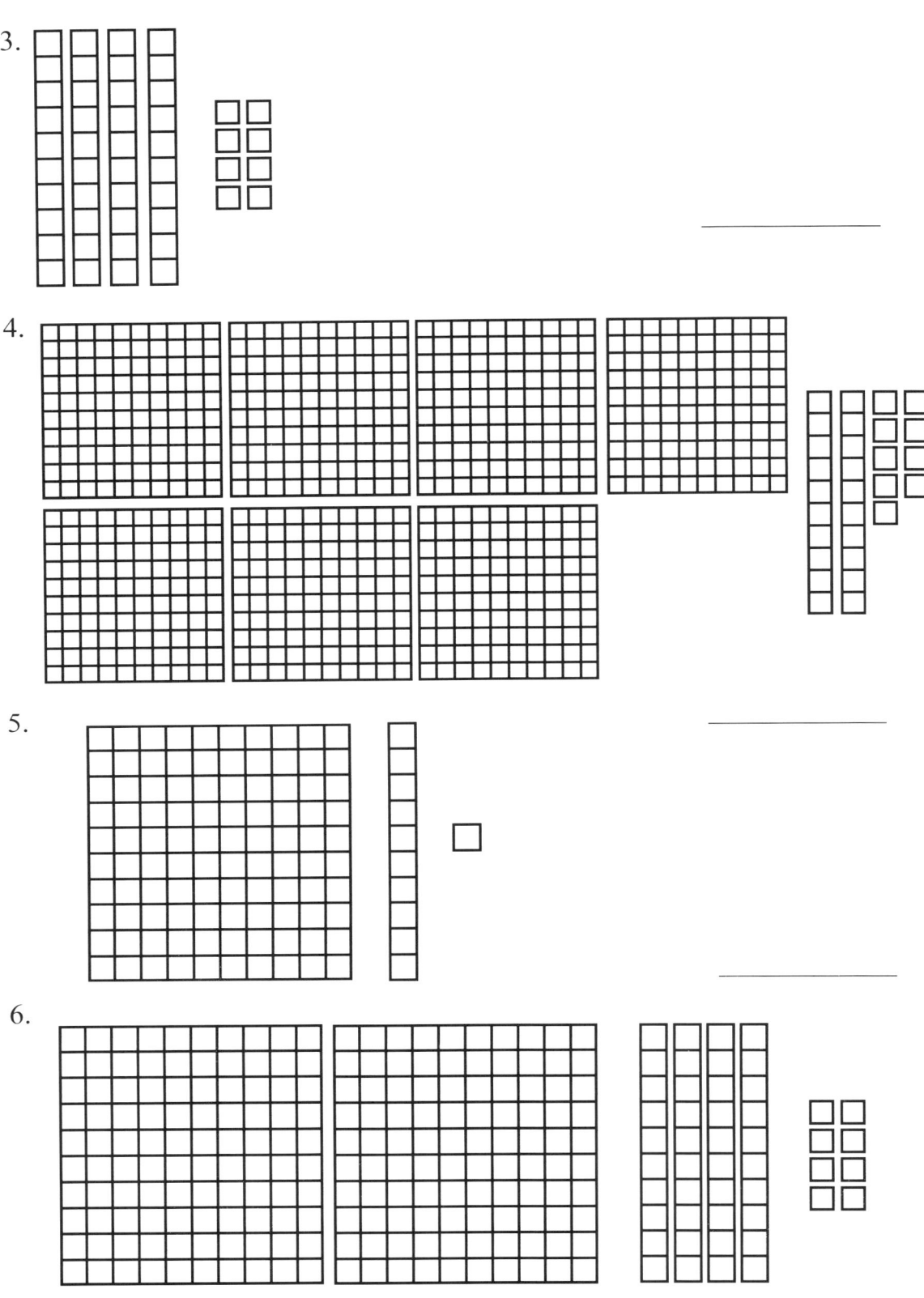

3. _____

4. _____

5. _____

6. _____

Draw models to show the numbers. (DOK 2)

7. 313

8. 25

9. 106

1.5 Place Value Mats (DOK 1,2)

Place value mats also model numbers. The place value is labeled in a chart. The number of circles stands for the digit in the given place value.

Example 1: Mrs. Finley used the place value mat below to show a number.

Thousands	Hundreds	Tens	Ones
OOO	OO	OOOOO	OOOO

Step 1: Count the O s to determine the digit for each place value.

Thousands	3
Hundreds	2
Tens	5
Ones	4

Step 2: Write the number.

Step 3: 3,254

Write the number that each place value mat represents. (DOK 1)

1.

Thousands	Hundreds	Tens	Ones
OO	OOOO	OOOOO	OO

_____

2.

Thousands	Hundreds	Tens	Ones
OOOOOOO	OOOOO	OOOOO	OO

_____

3.

Thousands	Hundreds	Tens	Ones
OOO	OOO	OOOO	OOOOO

_____

4.

Thousands	Hundreds	Tens	Ones
OOOO	OOOOOO	OOOOO	OOOOOO

_____

5.

Thousands	Hundreds	Tens	Ones
OOO	OO	OOOOO	OOOO

_____

6.

Thousands	Hundreds	Tens	Ones
OOOOOOO	OO	OOOOO	OOOOOO

_____

For numbers 7-13, choose the place value mat (A, B, C, or D) that gives the answer. (DOK 2)

A
Thousands	Hundreds	Tens	Ones

B
Thousands	Hundreds	Tens	Ones

C
Thousands	Hundreds	Tens	Ones

D
Thousands	Hundreds	Tens	Ones

7. Which place value mat shows a number with a digit in the ones place that is <u>two more</u> than the tens place? _____

8. Which place value mat shows a number with a digit in the thousands place that is <u>greater than</u> the tens place? _____

9. Which place value mat shows a number with the <u>same</u> digit in the thousands, hundreds, and tens place? _____

10. Which place value mat shows a number with a digit in the thousands place that is <u>smaller than</u> any other digit? _____

11. Which place value mat shows a number with a digit in the ones place that is <u>equal</u> to the digit in the thousands place? _____

12. Which place value mat shows a number with a digit in the hundreds place that is <u>equal</u> to the digit in the ones place? _____

13. Which place value mat shows a number with a digit in the ones place that is <u>equal</u> to three? _____

1.6 Comparing Numbers (DOK 2)

We can compare numbers to see which is bigger or smaller. Two symbols are used when comparing numbers, > and <. The open end of the symbol points to the larger number. The closed end of the symbol points to the smaller number.

Example 1: 10 > 5. This is read as 10 is greater than 5.

Example 2: 5 < 10. This is read as 5 is less than 10.

Example 3: Which place value shows that 4,330 is <u>bigger than</u> 4,130?

> **Step 1:** Look at the largest place value (thousands).
> 4 is in the thousands place of 4,330.
> 4 is in the thousands place of 4,130.
>
> Since the thousands place is the same in both numbers, look at the next place value.

> **Step 2:** Look at the next place value (hundreds).
> 3 is in the hundreds place of 4,330.
> 1 is in the hundreds place of 4,130.
> Since 3 is bigger than 1, the hundreds place value shows that 4,330 is bigger than 4,130, or 4,330 > 4,130.

Write the correct symbol, > or < , in the boxes provided. (DOK 2)

1. 26 ☐ 25

2. 12 ☐ 11

3. 95 ☐ 89

4. 24 ☐ 42

5. 5 ☐ 3

6. 19 ☐ 18

7. 5125 ☐ 5168

8. 2355 ☐ 2353

9.

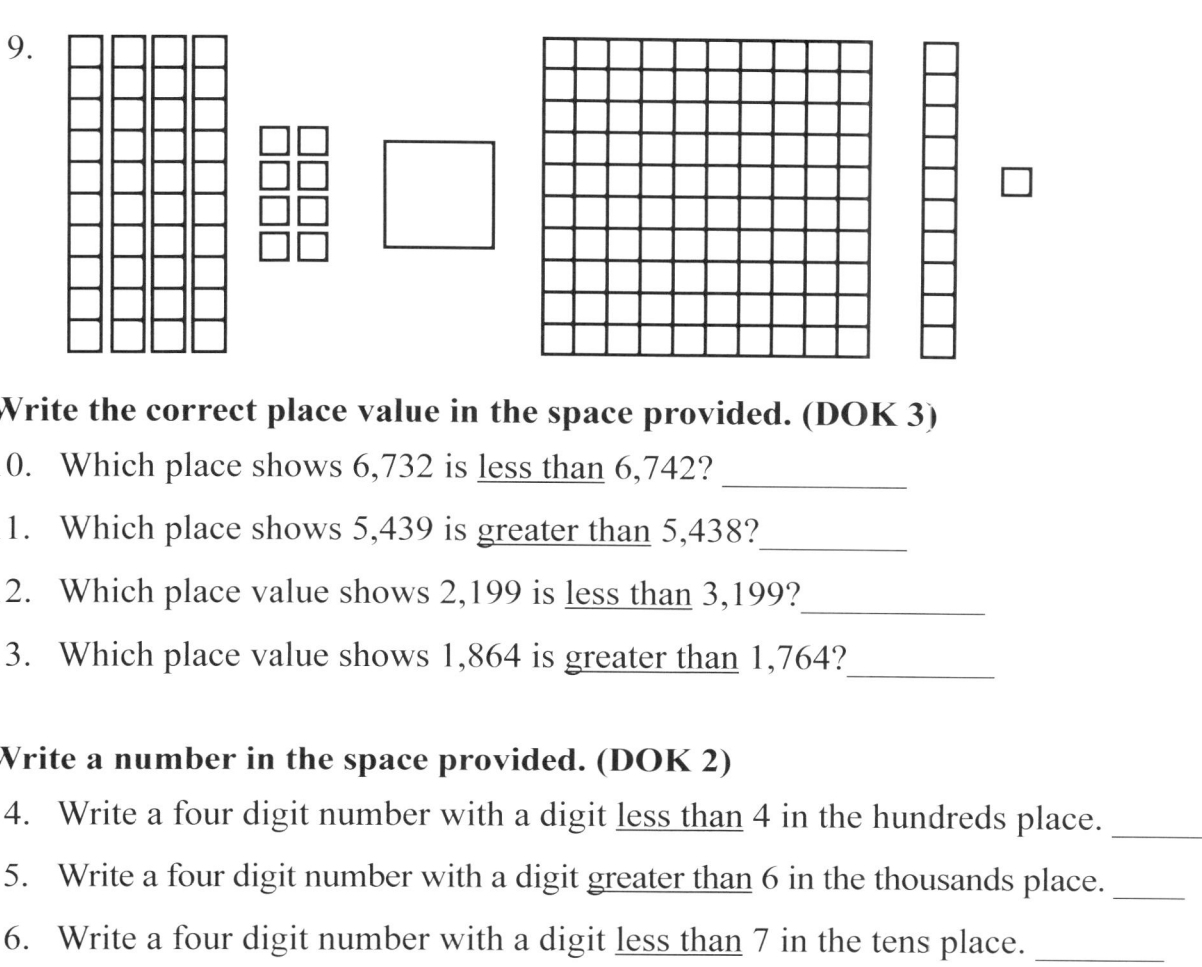

Write the correct place value in the space provided. (DOK 3)

10.   Which place shows 6,732 is <u>less than</u> 6,742? _____

11.   Which place shows 5,439 is <u>greater than</u> 5,438? _____

12.   Which place value shows 2,199 is <u>less than</u> 3,199? _____

13.   Which place value shows 1,864 is <u>greater than</u> 1,764? _____

Write a number in the space provided. (DOK 2)

14.   Write a four digit number with a digit <u>less than</u> 4 in the hundreds place. _____

15.   Write a four digit number with a digit <u>greater than</u> 6 in the thousands place. _____

16.   Write a four digit number with a digit <u>less than</u> 7 in the tens place. _____

17.   Write a four digit number with a digit <u>greater than</u> 3 in the ones place. _____

1.7 Ordering Numbers (DOK 1, 2)

To **order numbers**, sort them from smallest to largest or from largest to smallest. **Number lines** can be used to order numbers. The farther right on a number line, the larger the number. The farther left on a number line, the smaller the number. The closed end of the less than and greater than symbols always points to the smaller number when comparing two numbers: $2 < 4$ and $4 > 2$.

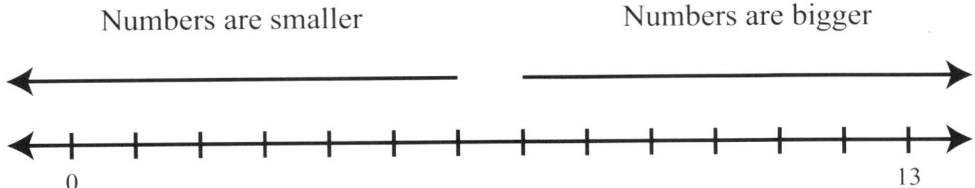

Numbers are smaller   Numbers are bigger

0   13

Example 1: Order the numbers from <u>smallest</u> to <u>largest</u>.
  57   3,842   17   623   1,542

  Step 1: Sort the numbers with 1 or 2 digits from smallest to largest.
  17   57

  Step 2: Sort the numbers with 3 digits from smallest to largest. Add them to the list from Step 1.
  623
  So now we have 17   57   623

  Step 3: Sort the numbers with 4 digits from smallest to largest. Add them to the list from Step 2.
  1,542   3,842

  Step 4: 17   57   623   1,542   3,842

Order the numbers from <u>smallest</u> to <u>largest</u>. (DOK 2)

1. 18   742   12   1,954   367   3. 243   27   3,647   3,646   23

   _____   _____

2. 336   6,333   633   36   63   4. 8,543   9,528   78   742   97

   _____   _____

Order the numbers from <u>largest</u> to <u>smallest</u>. (DOK 2)

5. 5,247   39   1,542   37   488

_____

6. 6,422   244   422   4,622   2

_____

7. 81   8,118   18   811   8   11

_____

8. 2,789   912   834   91   1,145

_____

(DOK 1, 2)

9. Which set of numbers is ordered <u>smallest</u> to <u>largest</u>?
    A) 3573, 3537, 3583
    B) 3753, 3754, 3781
    C) 3735, 3835, 3753

10. Which set of numbers is ordered <u>largest</u> to <u>smallest</u>?
    A) 1117, 1107, 1097
    B) 1097, 1117, 1107
    C) 1107, 1117, 1097

11. Which set of numbers is ordered from <u>smallest</u> to <u>largest</u>?
    A) 8972, 5643, 2331
    B) 1992, 3987, 2217
    C) 1223, 2471, 4852

12. Which set of numbers is ordered <u>largest</u> to <u>smallest</u>?
    A) 952, 1107, 1052
    B) 1107, 1052, 952
    C) 1107, 952, 1052

13. Which set of numbers is ordered <u>smallest</u> to <u>largest</u>?
    A) 5642, 5651, 5652
    B) 5652, 5642, 5651
    C) 5651, 5642, 5652

(DOK 3)

14. Explain why these numbers are in order from <u>largest</u> to <u>smallest</u>?

    7,621

    7,620

    7,520

    _____

15. Order the numbers. Write them in the correct place on the number line. Label the number line as needed. 18, 7, and 3.

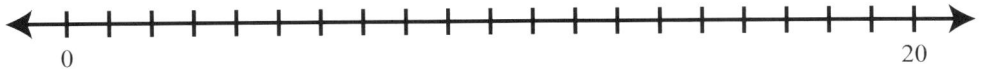

0   20

1.8 Rounding Numbers (DOK 1, 2)

Rounding numbers help estimate the value of a number. A rounded number is a simpler way to see a number. In this section, we will round numbers to the nearest ten, nearest hundred, and nearest thousand. Round up when a digit is greater than or equal to 5. Round down when a digit is less than 5.

Round to the nearest ten	Decide if a number is closer to 0 or 10, 10 or 20, 20 or 30, 30 or 40, and so on.
Round to the nearest hundred	Decide if a number is closer to 0 or 100, 100 or 200, 200 or 300, 300 or 400, and so on.
Round to the nearest thousand	Decide if a number is closer to 0 or 1000, 1000 or 2000, 2000 or 3000, 3000 or 4000, 4000 or 5000, and so on.

Special Case: Rounding with 9s

	Tens	**Hundreds**	**Thousands**
9	Closer to 0 or 10	Does not apply	Does not apply
99	Closer to 90 or 100	Closer to 0 or 100	Does not apply
999	Closer to 990 or 1,000	Closer to 900 or 1,000	Closer to 0 or 1,000
9999	Closer to 9990 or 10,000	Closer to 9,900 or 10,000	Closer to 9,000 or 10,000

Example 1: Round 32 to the nearest ten.

    Step 1: Look at the place value to the right of the tens place.
            3<u>2</u>

    Step 2: Ask if the underlined number in Step 1 is greater than or equal to 5. It is not, so round 32 down to 30.

Example 2: Round 472 to the nearest hundred.

    Step 1: Look at the place value to the right of the hundreds place.
            4<u>7</u>2

    Step 2: Ask if the underlined number in Step 1 is greater than or equal to 5. It is, so round 472 up to 500.

Example 3: Round 2,389 to the nearest thousand.

    Step 1: Look at the place value to the right of the thousands place.
            2,<u>3</u>89

    Step 2: Ask if the underlined number in step 1 is greater than or equal to 5. It is not, so round 2,389 down to 2,000.

Round to the nearest ten. (DOK 1)

1. 523 _____
2. 6,745 _____

3. 1,324 _____
4. 882 _____

Round to the nearest hundred. (DOK 1)

5. 659 _____
6. 4,478 _____

7. 107 _____
8. 2,731 _____

Round to the nearest thousand. (DOK 1)

9. 4,908 _____
10. 6,743 _____

11. 1,342 _____
12. 7,453 _____

Circle the correct answer. (DOK 2)

13. Which number is rounded to the nearest hundred? 8,750   9,010   8,800

14. Which number is rounded to the nearest ten? 6,349   6,350   6,305

15. Which number is rounded to the nearest thousand? 8,000   8,001   8,100

1.9 Number Sense Enrichment (DOK 1, 2, 3)

Read each problem and solve. (DOK 1)

1. Henry was counting a box of bolts. He counted 2,000 + 400 + 10 + 6 bolts. How many bolts were there in all? _____

2. Marcy counted 1,678 pennies in her piggy bank. Write the number of pennies in word form.

3. Enrico wrote a thank you note to his grandmother for the 1,755 stickers he got for his birthday. Write 1,755 in <u>expanded</u> form.

(DOK 2)

4. Mrs. Moreno asked the class to write the number from the model pictured below. What is the number?

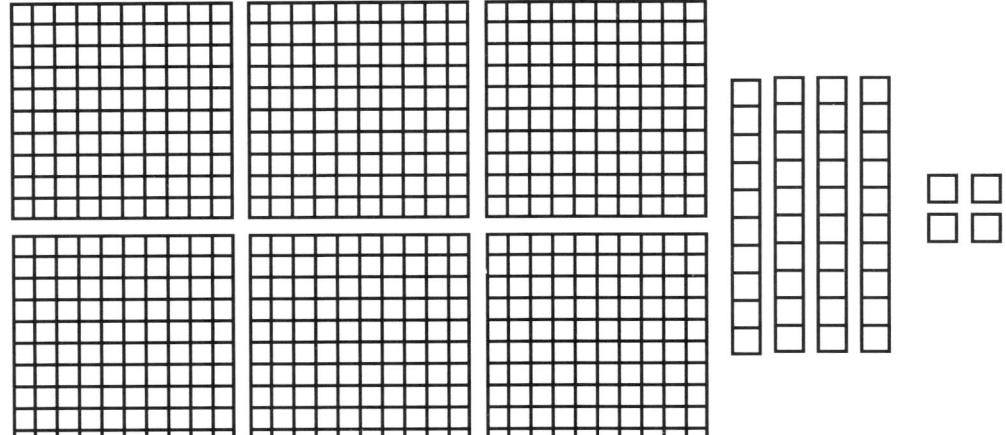

5. Alice has 67 nickels. Her sister, Abby, has 76 nickels. Who has more nickels? Write > or < .

Alice ☐ Abby

6. Jamal's mother asked him to order her checks. The check numbers were 1415, 1417, 1412, 1416, and 1418. Order the check numbers from <u>smallest</u> to largest.

_____

(DOK 3)

7. Noah rounded the number 9,562 to the nearest hundred. His answer is 9,500. Is Noah correct? If not, explain.

_____

8. Jennifer was asked to pick a four-digit number that has a tens place <u>greater than</u> 7. She chose the number 2,158. Is Jennifer correct? If not, explain.

_____

9. Explain why these 3 numbers are in correct order from <u>smallest</u> to <u>largest</u>.
   8,110   8,111   8,112

_____

10. Bernardo counted 482 stamps in his uncle's stamp collection. Write the number of stamps in expanded form.

_____

11. Which number, if rounded to the nearest hundred, would equal 5,500? Explain your reasoning.
   5,412   5,450   5,440

_____

Chapter 1 Review

Answer each question. (DOK 1)

1. Write the number 643 in word form.

    _____

2. Write the number 8,110 in word form.

    _____

3. Write nine thousand, nine hundred seventy-seven in <u>standard</u> form.

    _____

4. Write three thousand, one hundred twenty-two in <u>standard</u> form.

    _____

5. Write 2,712 in <u>expanded</u> form.

    _____

6. Write 6,336 in <u>expanded</u> form.

    _____

7. Write 3,000 + 500 + 60 + 8 in <u>standard</u> form. _____

8. Write 7000 + 800 + 2 in <u>standard</u> form. _____

9. What place value is the number 0 in 4,072? _____

10. What place value is the number 1 in 2,817? _____

11. What number is in the tens place in 6,789? _____

12. What number is in the hundreds place in 8,243? _____

13. What is the number value of the place value mat below?

Thousands	Hundreds	Tens	Ones
●●●	●●	●●●	●●
●●	●●	●●●	●

_____

(DOK 2)

14. Use the symbols > or < to fill in the boxes below.

    17 ☐ 20   68 ☐ 54

15. Put the list of numbers below in order from <u>smallest</u> to <u>largest</u>.
    84   211   5,417   684   212   5,415

    _____

16. Put the list of numbers below in order from <u>largest</u> to <u>smallest</u>.
    62   1,500   15   6,200   151   620

    _____

(DOK 1)

17. Round 3,781 to the nearest hundred. _____

18. Round 849 to the nearest hundred. _____

19. Round 783 to the nearest ten. _____

20. Round 4,567 to the nearest ten. _____

21. Round 1,560 to the nearest thousand. _____

22. Round 6,922 to the nearest thousand. _____

(DOK 2)

23. Which number is larger, 28 or 31? _____

24. Which number is smaller, 19 or 15? _____

(DOK 1)

25. What number is shown by the blocks below?

    _____

26. What number is shown by the blocks below?

_____

Read each problem and solve. (DOK 3)

27. Brandon was asked to pick a four-digit number that has a thousands place <u>less than</u> 4. He chose 6,442. Is Brandon correct? If not, explain.

28. Fred counted 168 ceiling tiles in the school library. Write the number of ceiling tiles in expanded form.

29. Which number, if rounded to the nearest ten, would equal 7,460? Explain your reasoning.

 7,462 7,465 7,466

For additional practice, please see Chapter 1 Test located in the Teacher Guide.

Chapter 2
Addition and Subtraction

This chapter covers the following Grade 3 standards:

	Content Standard
Number and Operations in Base 10	3.NBT.1, 3.NBT.2, 3.OA.9

2.1 Properties of Addition (DOK 1)

The table below lists the three basic number properties (or rules) for addition.

Property	Model	Rule
Identity	□ □ □ □ □ + 0 = □ □ □ □ □ 5 + 0 = 5	When adding two numbers, and one number is zero, the sum is equal to the other number.
Commutative	□ □ □ □ + □ □ = □ □ + □ □ □ □ 4 + 2 = 2 + 4	When adding two numbers the sum will remain the same no matter which order you use.
Associative	(□ + □ □) + □ □ □ = □ + (□ □ + □ □ □) (1 + 2) + 3 = 1 + (2 + 3)	When adding three or more numbers, the sum is the same no matter how you group them.

Look at each number sentence below. Write the number sentence that should be on the right side of the equal sign. Also, write the property used. You do not need to solve the problems. The first one is done for you. (DOK 1)

1. $6 + 9 = \underline{9 + 6}$ (Commutative)

2. $1 + (4 + 5) =$ _____

3. $7 + 0 =$ _____

4. $12 + 27 =$ _____

5. $18 + 0 =$ _____

6. $22 + (18 + 13) =$ _____

7. $39 + 14 =$ _____

8. $10 + (6 + 4) =$ _____

Use the Commutative Property to answer the following questions. (DOK 1)

9. $6 + 5 = 5 + \square$

10. $3 + 4 = \square + 3$

11. $6 + 17 = \square + 6$

12. $\square + 2 = 2 + 9$

13. $\square + 5 = 5 + 17$

14. $8 + \square = 4 + 8$

Use the Identity Property to answer the following questions. (DOK 1)

15. $63 + 0 = \square$

16. $0 + 5 = \square$

17. $9 + \square = 9$

Use the Associative Property to answer the following questions. (DOK 1)

18. $(4 + 2) + 5 = 4 + (2 + \square)$

19. $(6 + 7) + 34 = 6 + (7 + \square)$

20. $(1 + 8) + 9 = \square + (8 + 9)$

21. $(2 + \square) + 4 = 2 + (7 + 4)$

22. $(7 + 3) + 1 = 7 + (\square + 1)$

23. $(4 + 9) + \square = 4 + (9 + 2)$

2.2 Two-Digit Addition (DOK 1, 2)

$$②+①=\boxed{3}$$

Addends are numbers that are added together. The sum of the addends 2 and 1 is 3. Column addition is used when adding vertically.

Whether it is 2-digit addition or 3-digit addition, addition should be done between the digits with respect to their place values. The ones place digit should be added to the other ones place digit, the tens place digit should be added to the other tens place digit, and so on.

Regrouping occurs when the sum of the addends is larger than 9. If we are adding the digits in the ones place addends, and the sum is 10, 0 should be placed under the ones column, and 1 should be placed above the tens column. When adding the tens column, be sure to include the 1 that was carried over in regrouping. Once you reach the left most column, place the total sum below that column.

Example 1: Find the sum of 12 and 14.

Step 1: Always line up the columns by place value.

$$
\begin{array}{r}
\text{Tens } \text{Ones} \\
1\,|\,2 \\
+\ 1\,|\,4 \\
\hline
\end{array}
$$

Step 2: Add the digits in the ones place.

$$
\begin{array}{r}
\text{Tens } \text{Ones} \\
1\,|\,2 \\
+\ 1\,|\,4 \\
\hline
6
\end{array}
$$

Since 6 is less than 9, we do not need to regroup.

Step 3: Add the digits in the tens column.

$$
\begin{array}{c}
\overset{\text{Tens}\ \text{Ones}}{} \\
1\ 2 \\
+\ 1\ 4 \\
\hline
2\ 6
\end{array}
$$

Answer: 26

Example 2: Caroline has 22 stickers in her album. Her mom gives her 18 more stickers. How many stickers does Caroline have in all?

Step 1: Find the sum of 22 and 18.

Step 2: Line up the columns by place value.

$$
\begin{array}{c}
\overset{\text{Tens}\ \text{Ones}}{} \\
2\ 2 \\
+\ 1\ 8 \\
\hline
\end{array}
$$

Step 3: Add the digits in the ones place.
2 + 8 = 10
Since 10 is greater than 9, regroup.

Step 4: Regroup.

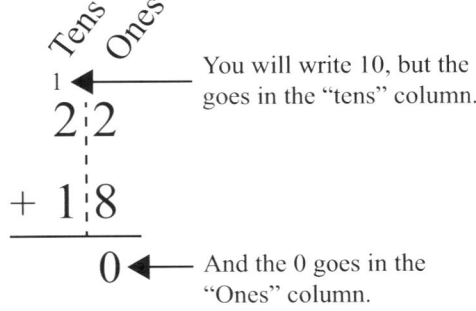

You will write 10, but the 1 goes in the "tens" column.

And the 0 goes in the "Ones" column.

Step 5: Add the digits in the tens column. Don't forget to add the 1 that was carried over with regrouping.

```
     Tens  Ones
       1
       2 : 2
     + 1 : 8
     ─────────
       4   0
```

Answer: Caroline has 40 stickers in all.

Find the sum. (DOK 1)

1.   72
    + 37

2.   55
    + 96

3.   90
    + 43

4.   19
    + 65

5.   44
    + 12

6.   71
    + 13

7.   99
    + 11

8.   17
    + 71

9.   84
    + 37

10. 22
    + 24

11. 11
    + 29

12. 34
    + 12

13. 55
    + 76

14. 12
    + 68

15. 45
    + 32

2.3 Three-Digit Addition (DOK 1, 2)

Three-digit addition is the same as two-digit addition, but we need to add another column for the hundreds place.

Example 1: Find the sum: $627 + 123$

Step 1: Always line up the place value columns.

$$
\begin{array}{c c c}
\text{Hundreds} & \text{Tens} & \text{Ones} \\
6 & 2 & 7 \\
+\ 1 & 2 & 3 \\
\hline
\end{array}
$$

Step 2: Add the digits in the ones place.

$$
\begin{array}{c c c}
 & 1 & \\
6 & 2 & 7 \\
+\ 1 & 2 & 3 \\
\hline
 & & 0 \\
\end{array}
$$

$7 + 3 = 10$ ones or 1 ten and 0 ones. Regroup by carrying 1 ten in the tens column and 0 in the ones column.

Step 3: Add the digits in the tens column. Add the 1 that was carried over from regrouping.

$$
\begin{array}{c c c}
 & 1 & \\
6 & \mathbf{2} & 7 \\
+\ 1 & \mathbf{2} & 3 \\
\hline
 & 5 & 0 \\
\end{array}
$$

Step 4: Add the digits in the hundreds column.

$$
\begin{array}{r}
\overset{\scriptstyle 1}{6}\;2\;7 \\
+\,1\;2\;3 \\
\hline
7\;5\;0
\end{array}
$$

Answer: 750

Example 2: Find the sum of 182 and 194.

Step 1: Line up the place value columns.

$$
\begin{array}{r}
1\;8\;2 \\
+\,1\;9\;4 \\
\hline
\end{array}
$$

Step 2: Add the digits in the ones place.

$$
\begin{array}{r}
1\;8\;2 \\
+\,1\;9\;4 \\
\hline
6
\end{array}
$$

Since 6 is less than 9, do not regroup.

Step 3:   Add the digits in the tens place.

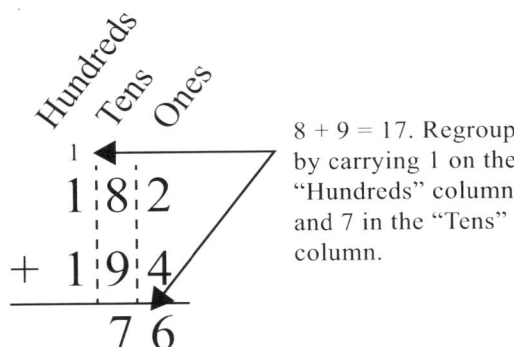

8 + 9 = 17. Regroup by carrying 1 on the "Hundreds" column and 7 in the "Tens" column.

Step 4:   Add the digits in the hundreds place. Don't forget to add the 1 from regrouping.

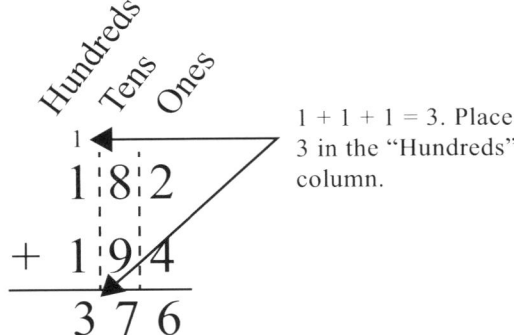

1 + 1 + 1 = 3. Place 3 in the "Hundreds" column.

Answer: 376

Example 3:   Find the sum of 586 and 796.

Step 1:   Line up the place value columns.

$$
\begin{array}{r}
5\ 8\ 6 \\
+\ 7\ 9\ 6 \\
\hline
\end{array}
$$

Step 2:   Add the digits in the ones place.

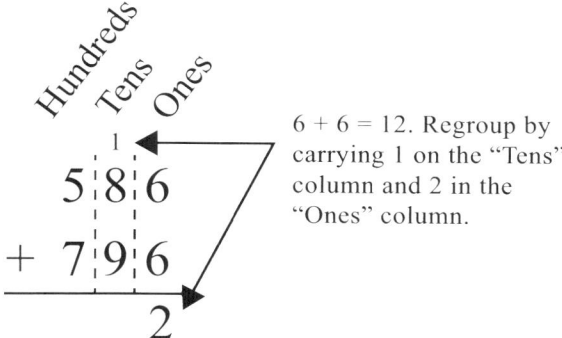

6 + 6 = 12. Regroup by carrying 1 on the "Tens" column and 2 in the "Ones" column.

Step 3:   Add the digits in the tens place. Don't forget to add the 1 that you regrouped from the ones place.

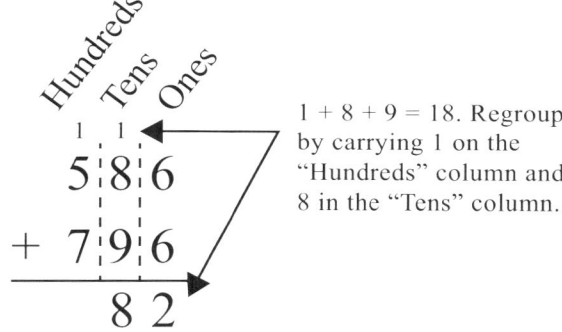

1 + 8 + 9 = 18. Regroup by carrying 1 on the "Hundreds" column and 8 in the "Tens" column.

Step 4: Add the digits in the hundreds place. Don't forget to add the 1 from regrouping in the tens place.

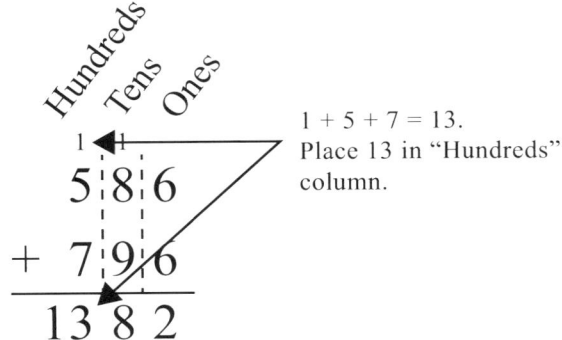

$1 + 5 + 7 = 13$.
Place 13 in "Hundreds" column.

Answer: 1382

Find the sum. (DOK 1, 2)

1. 248
    + 672

2. 113
    + 214

3. 110
    + 100

4. 678
    + 112

5. 326
    + 543

6. 148
    + 233

7. 417
    + 429

8. 104
    + 205

9. 748
    + 206

10. 214
    + 195

11. 118
    + 392

12. 285
    + 408

13. 912
    + 571

14. 257
    + 487

15. 689
    + 124

2.4 Applying Addition (DOK 2)

Words Indicating Addition	Example	Add
and	6 **and** 8	6 + 8
increased	A $15 shirt is **increased** by $5.	$15 + $5
more	Kate has 3 coins. Kate's mom has 8 **more** coins than Kate.	3 + 8
more than	Josh has 10 points. Will has 5 **more than** Josh.	10 + 5
plus	8 baseballs **plus** 4 baseballs	8 + 4
sum	the **sum** of 3 and 5	3 + 5
total	the **total** of 10, 14, and 15	10 + 14 + 15
all together	There are 5 cats and 3 dogs How many pets are there **all together**?	5 + 3
in all	Ted has 3 red marbles, 6 blue marbles, and 2 white marbles How many marbles are there **in all**?	3 + 6 + 2

Follow these steps to solve word problems:

Step 1: Read the problem carefully.

Step 2: Look for words that show the operation.

Step 3: Solve.

Example 1: If Bailey has 5 cupcakes and Parker has 4, how many cupcakes do they have all together?

5 cupcakes + 4 cupcakes = 9 cupcakes

Use a number sentence to solve.
$5 + 4 = 9$

Answer: 9 cupcakes

Solve the following problems. Show your work by writing a number sentence. (DOK 2)

1. Wayne has 4 basketballs and 6 baseballs. How many balls does Wayne have all together? _____

2. Katie picked 212 apples. Monica picked 62 more apples than Katie. How many apples did Monica pick? _____

3. In Debra's room, there are 4 books on the bed and 9 on the floor. How many books are in Debra's room? _____

4. There are 24 students in Miss Gallagher's 3rd grade class and 27 students in Mrs. Bentley's 3rd grade class. How many students are there in the two classes? _____

5. The Happy Pet Store has 14 kittens, 12 puppies, and 37 hamsters to sell. How many pets are there in all? _____

6. Mia counted 12 birds on the first bird feeder, 17 birds on the second bird feeder, and 11 birds on the third bird feeder. How many birds did Mia count in all? _____

7. Matt has 14 toy cars in his bedroom, 12 in the living room, and 6 in his playroom. How many toy cars does Matt have in all? _____

Use the chart to answer the next 5 questions. (DOK 3)

8. How many cans did both Joan and Mike collect in all? _____

9. How many total cans did the girls collect? _____

10. How many cans did Russell and Ted collect all together? _____

11. How many cans did the boys collect all together? _____

12. How many total cans did Mrs. Bing's class collect? _____

Mrs. Bing's Class Can-A-Thon Results	
Name	**Cans Collected**
Russell	5
Mike	6
Ted	8
Penny	10
Joan	6

2.5 Two-Digit Subtraction (DOK 1, 2)

$$⑨ - ④ = \boxed{5}$$

The difference of 9 and 4 is 5. Subtract the smaller number (subtrahend) from the larger number (minuend) to find the difference.

Column subtraction is used when subtracting vertically. The larger number goes on top. Subtraction should start from the right and end to the left.

Whether it is two-digit subtraction or three-digit subtraction, subtraction between digits is completed with respect to their place values. The digit in the ones place should be subtracted from the digit above in the ones place. The same is to be done in the tens place as well.

Regrouping in subtraction occurs when the digit in the bottom number is greater than the digit in the top number.

Example 1: Noah has 62 toys. He gave 13 to his cousin. How many toys does Noah have left?

Step 1: Find the difference. Line up the columns. The larger number listed goes on the top.

$$\begin{array}{r} 6\,2 \\ -\ 1\,3 \\ \hline \end{array}$$

Step 2: Start in the ones columns and subtract the digit on the bottom from the digit on the top.

$$\begin{array}{r} \text{Tens} \ \text{Ones} \\ 6\,2 \\ -\ 1\,3 \\ \hline \end{array}$$

Since 3 is larger than 2, you will need to regroup (borrow).

Step 3:   Regroup. Borrow 10 from the tens column.

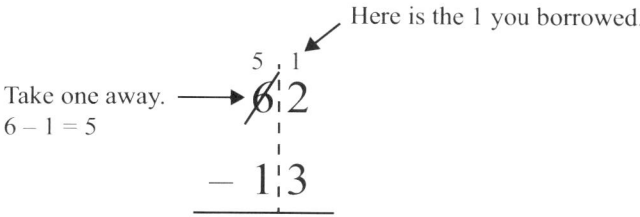

Here is the 1 you borrowed.

Take one away.
6 – 1 = 5

$$\begin{array}{r} \overset{5}{\cancel{6}}\overset{1}{2} \\ - 1\,3 \\ \hline \end{array}$$

Step 4:   Now go back and subtract 3 from 12.

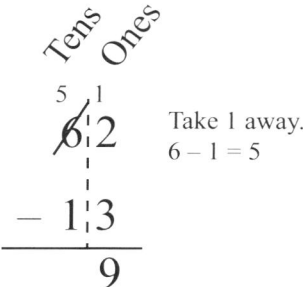

Take 1 away.
6 – 1 = 5

$$\begin{array}{r} \overset{5}{\cancel{6}}\overset{1}{2} \\ - 1\,3 \\ \hline 9 \end{array}$$

Step 5:   Subtract the digits in the tens place.

Take 1 away.
6 – 1 = 5

$$\begin{array}{r} \overset{5}{\cancel{6}}\overset{1}{2} \\ - 1\,3 \\ \hline 4\,9 \end{array}$$

Answer: Noah has 49 toy cars left.

Example 2:   Find the difference between 90 and 78.

Step 1:   Line up the columns.

$$\begin{array}{r} 9\,0 \\ - 7\,8 \\ \hline \end{array}$$

Step 2:   Subtract the digits in the ones place.
8 is greater than 0, so regroup.

Step 3: Regroup. Borrow 10 from the tens column.

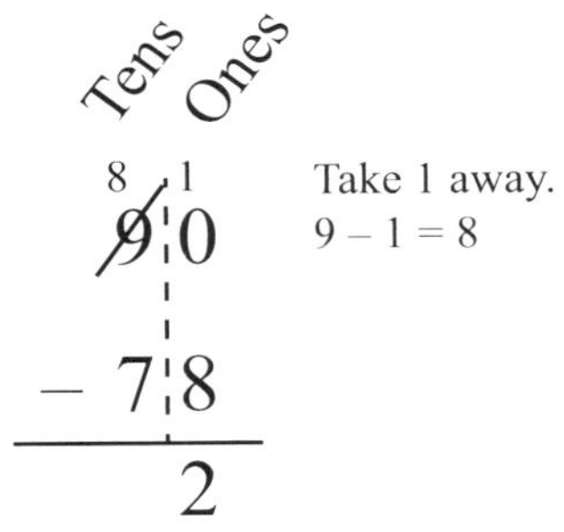

Tens Ones

$\overset{8}{\cancel{9}}$ | $\overset{1}{0}$ Take 1 away.
            $9 - 1 = 8$

$- \ 7 | 8$

2

$10 - 8 = 2$

Step 4: Subtract the digits in the tens column. Since we borrowed 10 from the tens place, the 9 becomes an 8.

$8 - 7 = 1$

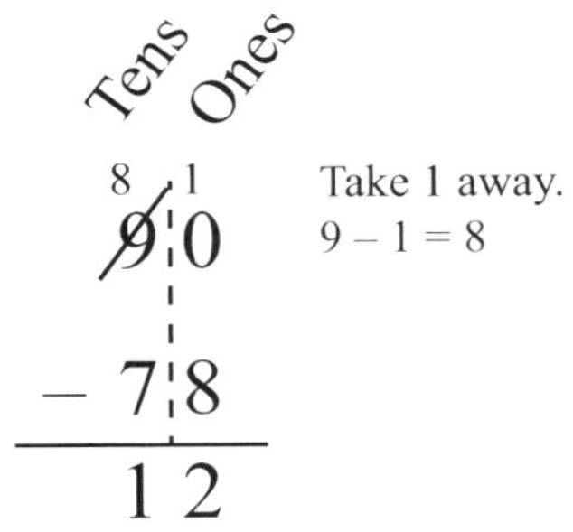

Tens Ones

$\overset{8}{\cancel{9}}$ | $\overset{1}{0}$ Take 1 away.
            $9 - 1 = 8$

$- \ 7 | 8$

$1 \ 2$

Answer: 12

Find the difference. (DOK 1)

1. 92
    − 18

2. 60
    − 52

3. 56
    − 19

4. 87
    − 9

5. 99
    − 61

6. 50
    − 27

7. 56
    − 12

8. 71
    − 17

9. 79
    − 68

10. 25
     − 0

2.6 Three Digit Subtraction (DOK 1, 2)

Jack has 257 quarters and gives 169 to a charity. How many quarters does Jack have left?

Example 1: Find the difference: 257 − 169

Step 1: Line up the columns.

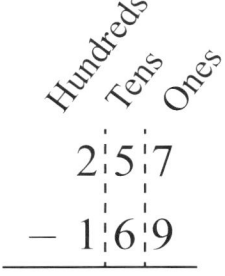

Step 2: Subtract the digits in the ones place.
9 is greater than 7, so regroup.

Step 3: Regroup. Borrow 10 from the tens column.
$17 - 9 = 8$

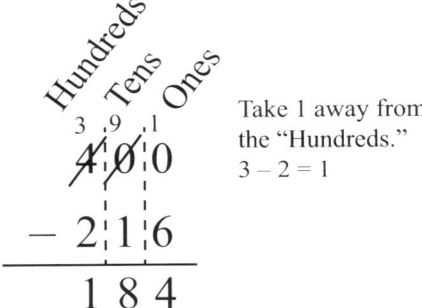

Take 1 away from
the "Tens."
$5 - 1 = 4$

Step 4: Subtract the digits in the tens column.
Since we borrowed 10 from the tens place, 5 becomes 4.
6 is greater than 4, so regroup.

Step 5: Subtract the numbers in the hundreds column.

Take 1 away from
the "Hundreds."
$3 - 2 = 1$

Since we borrowed 10 from the hundreds place, 2 becomes 1.

Answer: Jack has 88 quarters.

Example 2: Find the difference: $400 - 216$

Step 1: Line up the columns.

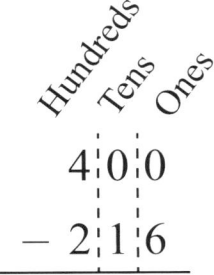

Step 2: Subtract the digits in the ones column. Since 6 is greater than 0, re-group.

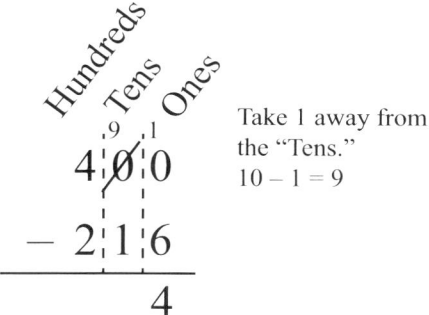

Take 1 away from the "Tens."
$10 - 1 = 9$

Step 3: Regroup. Borrow 10 from the tens column. Since there is a 0 in the tens column, borrow from the hundreds column and move it to the tens column. Now borrow from the tens column.
$10 - 6 = 4$

Step 4: Subtract the digits in the tens column. Since we borrowed 10 from the tens place, 0 becomes 9 in the tens place. Regrouping is done.
$9 - 1 = 8$

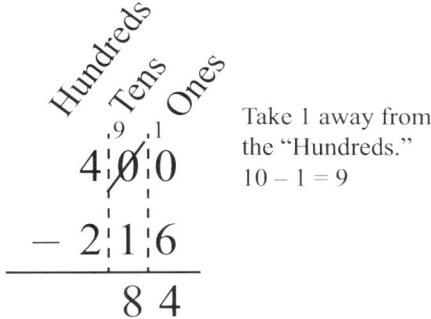

Take 1 away from the "Hundreds."
$10 - 1 = 9$

Step 5: Subtract the digits in the hundreds column. Since we borrowed from the hundreds place, 4 becomes 3.
$3 - 2 = 1$

Take 1 away from the "Hundreds."
$3 - 2 = 1$

Answer: $400 - 216 = 184$

Find the difference. (DOK 1)

1. 891
    $-$ 640

2. 207
    $-$ 138

3. 367
    $-$ 142

4. 751
    $-$ 137

5. 204
    $-$ 102

6. 504
    $-$ 315

7. 101
    $-$ 18

8. 217
    $-$ 123

9. 908
    $-$ 467

10. 224
    $-$ 173

11. 181
    $-$ 62

12. 827
    $-$ 99

2.7 Applying Subtraction (DOK 1, 2)

Words Indicating Subtraction	Example	Subtract
decreased	$16 **decreased** by $5	$16 – $5
difference	the **difference** between 18 and 6	18 – 6
less	14 days **less** 5	14 – 5
less than	Jose completed 2 laps **less than** Mike's 9	9 – 2
left	Ray sold 15 out of 35 tickets. How many did he have **left**?	35 – 15
lower than	This month's rainfall is 2 inches **lower than** last month's rainfall of 8 inches.	8 – 2
minus	15 **minus** 6	15 – 6
remain	Gary ate 4 of 12 cookies. How many **remain**?	12 – 4

*In subtraction word problems, you cannot always subtract the numbers in the order that they appear in the problem. Subtract the smaller number from the larger number.

Follow these steps to solve word problems:

Step 1: Read the problem carefully.

Step 2: Find the words indicating subtraction.

Step 3: Solve the problem.

Example 3: Nick has 7 bananas. He gives Patty 3 of them. How many bananas remain?
"Remain" tells us that we should subtract.

7 bananas – 3 bananas = 4 bananas

Answer: 4 bananas

Solve the word problems. Show your work by writing a number sentence. (DOK 2)

1. At the local zoo, John saw 6 tigers and 2 pandas. How many more tigers than pandas did he see?

   _____

2. Jacob ran a foot race in 87 minutes. His friend, Matthew, ran the same race in 78 minutes. By how many minutes was Matthew faster than Jacob?

   _____

3. Olivia has to sell 200 boxes of cookies for her troop. She sold 32 boxes on the first day and 54 boxes on the second day. How many boxes of cookies remain for Olivia to sell?

   _____

4. Miss Tiller bought 114 small candies to share with her 3rd grade class. The students ate 95 of the candies. How many candies remain?

   _____

5. Samantha has 763 pennies. She spends 212 of them on a toy and 89 of them on a bottle of bubbles. How many pennies does she have left?

   _____

6. Christopher has 54 marbles. He gave 22 of them to his younger brother, and his kitten batted 3 of them down the basement stairs. How many marbles remain for Christopher?

   _____

7. Madison made 14 hot pads to give to family as gifts. She gave 3 to her grandma, 4 to her mom, and 6 to her aunts. How many hot pads does she have left?

   _____

8. Flopsy the Clown starts the day with 125 balloons. He sells 27 of the balloons, gives away 7 of the balloons, and loses 3. How many balloons does Flopsy the Clown have left?

    _____

9. Ethan bought a package of 100 bean seeds. He planted 72 of the seeds. How many bean seeds does Ethan have left?

    _____

10. Gayle's doll is 18 inches tall. Sara's doll is 4 inches shorter than Gayle's. How tall is Sara's doll?

    _____

2.8 Number Sentences (DOK 1, 2)

In the number sentences $3 + 2 = 5$ and $2 + 3 = 5$, 3 and 2 are addends, and 5 is the sum.

In the number sentences $5 - 2 = 3$ and $5 - 3 = 2$, subtract each addend from the sum.

This explains that when we subtract one of the addends from the sum, we get the other addend.

Subtraction is the opposite of addition. To find the missing number in a subtraction sentence, use addition. To find the missing number in addition, use subtraction.

An **open number sentence** has a number missing. The missing number may be represented by a shape or letter, like a square, circle, or N.

Example 1: Find the missing value.
$$\square + 14 = 22$$

Step 1: We want to find the missing number that is to be added to 14 to get 22. Subtract 14 from 22.
$22 - 14 = 8$

Step 2: Check your work by replacing the square in the problem with the answer from Step 2.
$8 + 14 = 22$
This is correct.

Answer: The missing value is 8.

Example 2: The circle in the problem below represents a missing number. Find the missing number.
$54 - \bigcirc = 31$

   Step 1: Find the missing value by subtracting 31 from 54.
$54 - 31 = 23$

   Step 2: Check your work by replacing the circle with the answer from Step 1.
$54 - 23 = 31$
This is correct.

   Answer: The missing value is 23.

Find the missing value. (DOK 2)

1. $\Box + 7 = 11$

2. $501 + \bigcirc = 512$

3. $\bigcirc + 16 = 68$

4. $N + 315 = 435$

5. $22 + \Box = 57$

6. $61 + N = 83$

7. $\bigcirc + 362 = 715$

8. $N + 14 = 35$

9. $110 + N = 219$

10. $\Box + 31 = 74$

11. $67 - N = 15$

12. $18 - \Box = 3$

13. $27 - 14 = \bigcirc$

14. $482 - \Box = 461$

15. $112 - \bigcirc = 81$

16. $248 - N = 202$

17. $58 - \Box = 13$

18. $156 - \bigcirc = 44$

19. $78 - 25 = \Box$

20. $36 - 15 = \bigcirc$

2.9 Estimating (DOK 2)

Estimation is a way to quickly add or subtract large numbers. Rounding numbers is a way to estimate. Estimation never gives an exact answer.

Example 1: Estimate the sum: $317 + 596$.

Estimation

A) Find the tens place for both numbers. If it is less than 5, keep the number in the hundreds place the same. If it is 5 or more, round up to the next number.

B) For 317, 1 is less than 5, so we round down to 300.

C) For 596, 9 is greater than 5, so we round up to the next number, 600.

D) Add the rounded numbers. $300 + 600 = 900$. You may also use mental math to add. $3 + 6 = 9$. Then add 2 zeros. $317 + 596$ is around 900.

Compatible Numbers

A) Check the digit in the ones place. If the digit in the ones place is 3 or greater, round it to 5. If it is less than 3, round it to 0. If the digit in the ones place is 8 or greater, round it to 0 and increase the tens place by 1. If it is less than 8, round it to 5

B) For 317, 7 is less than 8, so round 317 to 315.

C) For 596, 6 is less than 8, so round down to the nearest compatible number, 595.

D) Add $315 + 595 = 910$.

Example 2: Estimate the difference: $62 - 48$.

Estimation

   A) Find the ones place for the first number. If it is less than 5, keep the number in the ones place the same. If it is 5 or more, round up to the next number.

   B) For 62, 2 is less than 5. Round down to 60.

   C) For 48, 8 is more than 5. Round up to the next number, 50.

   D) Subtract.
     $60 - 50 = 10$. You may also use mental math to subtracct. $6 - 5 = 1$. Then add 1 zero.

Answer: $60 - 50$ is around 10.

Compatible Numbers

   A) Check the digit in the ones place. If the digit in the ones place is 3 or greater, round it to 5. If it is less than 3, round it to 0. If the digit in the ones place is 8 or greater, round it to 0, and increase the tens place by 1. If it is less than 8, round it to 5.

   B) For 62, since 2 is less than 3, round it to 0, which is 60.

   C) For 48, since 8 is rounded to 0, and the tens place increases by 1, round up to 50.

   D) Subtract.
     $60 - 50 = 10$

Answer: $60 - 50$ is around 10.

Estimate the following. (DOK 2)

1. $69 - 55 = $ _____

2. $28 - 25 = $ _____

3. $44 + 23 = $ _____

4. $53 - 17 = $ _____

5. $98 + 45 = $ _____

6. $509 - 321 = $ _____

7. $948 + 356 = $ _____

8. $767 - 639 = $ _____

9. $153 + 95 = $ _____

10. $287 + 225 = $ _____

11. Julie has $4.75. She buys a piece of candy for $3.25. Estimate how much money Julie has after she pays for the candy.

12. Dan has 57 baseball cards. He sold 12 to his friend. Estimate how many baseball cards Dan has left.

2.10 Patterns Using Addition and Subtraction (DOK 2)

A group of numbers that increases or decreases by the same amount makes a pattern. A **pattern** follows a certain rule or order.

Finding the missing number in a pattern:

Step 1: Decide if the numbers are getting bigger or smaller

Step 2: Find the difference between each number.

Step 3: Use the answer from Step 2 to find the next number in the pattern.

Example 1: Find the next number in the pattern.
2, 5, 8, ____

Step 1: In this pattern, the numbers are increasing.

Step 2: The difference between each number is 3.
$2 + 3 = 5$
$5 + 3 = 8$

Step 3: Add 3 to the last number in the pattern.
$8 + 3 = 11$

Answer: The next number in the pattern is 11.

Example 2: Find the next number in the pattern.
12, 8, 4, _____

Step 1: In this pattern, the numbers are decreasing.

Step 2: To get from one number to the next, subtract by 4. So subtract 4 from the last number in the pattern.
$4 - 4 = 0$

Answer: 0 is the next number in the pattern.

Each row and column of the 100s chart shows a pattern with a certain rule.

The numbers 5, 15, 20, 25, 30, 35, 40, 45, 50, 55, 60, 65, 70, 75, 80, 85, 90, 95, and 100 follow the pattern of adding by **5.**

The numbers 100, 90, 80, 70, 60, 50, 40, 30, 20, and 10 follow the pattern of subtracting by **10.**

100s Chart

1	2	3	4	**5**	6	7	8	9	**10**
11	12	13	14	**15**	16	17	18	19	**20**
21	22	23	24	**25**	26	27	28	29	**30**
31	32	33	34	**35**	36	37	38	39	**40**
41	42	43	44	**45**	46	47	48	49	**50**
51	52	53	54	**55**	56	57	58	59	**60**
61	62	63	64	**65**	66	67	68	69	**70**
71	72	73	74	**75**	76	77	78	79	**80**
81	82	83	84	**85**	86	87	88	89	**90**
91	92	93	94	**95**	96	97	98	99	**100**

In the addition chart below, add the number in the top row with a number in the left most column. The cell where the two numbers meet is the sum.

Rules when adding:

Even + Even = Even   Even + Odd = Odd   Odd + Odd = Even

		Odd									
	+	**1**	**2**	**3**	**4**	**5**	**6**	**7**	**8**	**9**	**10**
Odd	**1**	2 Even	3	4	5	6	7	8	9	10	11
	2	3	4	5	6	7	8	9	10	11	12
	3	4	5	6	7	8	9	10	11	12	13
	4	5	6	7	8	9	10	11	12	13	14
	5	6	7	8	9	10	11	12	13	14	15
	6	7	8	9	10	11	12	13	14	15	16
	7	8	9	10	11	12	13	14	15	16	17
	8	9	10	11	12	13	14	15	16	17	18
	9	10	11	12	13	14	15	16	17	18	19
	10	11	12	13	14	15	16	17	18	19	20

Find the missing term(s). (DOK 2)

1. 4, 8, 12, _____, 20

2. 16, 13, 10, _____, _____

3. 12, 11, 10, 9, _____

4. 16, 11, 6, _____

5. 4, 10, 16, 22, _____

6. 2, 9, 16, _____, _____

7. 12, 16, 20, 24, _____

8. 1, 11, 21, 31, _____

9. 24, 20, 16, 12, _____

10. 3, 6, 9, 12, _____

11. 10, 15, 20, 25, _____

12. 14, 12, 10, 8, _____

13. 1, 9, 17, 25, _____

14. 76, 77, 78, 79, _____

15. 34, 36, 38, 40, _____

16. 60, 54, 48, 42, _____

17. Ashley put 3 cents in her piggy bank. She added 3 cents to the piggy bank each day for the next 4 days. Fill in the chart to see how many pennies she has now.

Day 1	Day 2	Day 3	Day 4	Day 5
3				

18. Marco got 2 toys each from his father, his uncle John, his uncle Barry, and his cousin Frederico for his birthday. Fill in the chart below to see how many toys Marco got for his birthday.

	Marco's Father	Uncle John	Uncle Barry	Frederico
Toys Given	2			
Total No. of Toys				

2.11 Adding and Subtracting Enrichment (DOK 3)

Solve the addition problems below by drawing a diagram to show your work. The first one is done for you. (DOK 3)

1. 13 + 5 = ☐

   ☐☐☐☐☐☐☐☐☐☐ ☐☐☐ + ☐☐☐☐☐

   = 18

2. 12 + 3 = ☐

3. 10 + 7 = ☐ .

The problems below include both addition and subtraction in the same problem. Show your work by writing number sentences. (DOK 3)

4. Heather bought 4 apples. When she came home, she found her mother already had 17 apples in the pantry. Heather and her two brothers ate 3 of the apples. How many apples remain?

5. Andrew has 854 pennies in his penny box. He gives 158 of the pennies to his younger sister. His father gives him 12 more pennies that evening. How many pennies does Andrew have now?

6. At the beginning of the school year, Mrs. David's class had 32 students. There were not enough desks in Mrs. David's class, so 5 of the students went to another class where there was room for them. A month later, 2 of her students moved away. How many students are in her class now?

7. Mr. Bryant picked 37 tomatoes from his garden early in the morning. He and his family ate 14 of the tomatoes at lunch. Later in the day, Mr. Bryant picked another 8 tomatoes. How many tomatoes does Mr. Bryant have now?

8. Gina has 52 dollar bills. She deposits 20 of them into her savings account. She spends 2 of the dollars on a new toy. Her uncle gives her 5 more dollars for washing his car. How many dollar bills does Gina have at home now?

9. Taylor has 110 seashells in his collection. He finds 9 more seashells on the beach today. He gives 42 of the seashells to his mother for her art project. How many seashells does Taylor have now?

10. Olivia and her mother go to a pet store. When they walk in the store, there are 16 puppies for sale. While they are there, 2 of the puppies are sold. A breeder brings in 5 more puppies. How many puppies are there in the store now?

11. When Joseph got up for school this morning, it was 56° Fahrenheit outside. By 3:00 p.m., it was 22 degrees warmer. When Joseph went to bed, it was 15 degrees cooler than it was at 3:00 p.m. What was the temperature when Joseph went to bed?

12. Daniel's dog has 3 pounds of dog food left. Daniel's mother buys a 10-pound bag of dog food. Daniel's dog eats 1 pound of the food during the day. Daniel's father buys another 20 pounds of dog food on his way home from work. How much dog food is there in the house at the end of the day?

13. Emily owns 15 books. She checks out 2 more books from the library. She gives her mother 4 books of her own to sell at their yard sale. How many books does Emily have in all now, including the books from the library?

Chapter 2 Review

Find the sum. Regroup if needed. (DOK 1 and 2)

1. $\begin{array}{r} 27 \\ + 82 \\ \hline \end{array}$
2. $\begin{array}{r} 322 \\ + 145 \\ \hline \end{array}$
3. $\begin{array}{r} 455 \\ + 225 \\ \hline \end{array}$
4. $\begin{array}{r} 14 \\ + 41 \\ \hline \end{array}$
5. $\begin{array}{r} 204 \\ - 102 \\ \hline \end{array}$

Find the difference. Regroup if needed. (DOK 1 and 2)

6. $\begin{array}{r} 87 \\ - 12 \\ \hline \end{array}$
7. $\begin{array}{r} 788 \\ - 222 \\ \hline \end{array}$
8. $\begin{array}{r} 444 \\ - 222 \\ \hline \end{array}$
9. $\begin{array}{r} 732 \\ - 156 \\ \hline \end{array}$
10. $\begin{array}{r} 54 \\ - 31 \\ \hline \end{array}$

Solve the addition problem below by drawing a diagram to show your work. (DOK 3)

11. $16 + 2 = \boxed{}$

Read each problem. Answer the question. (DOK 2)

12. $\bigcirc + 147 = 298$

13. Find the missing number: 10, 12, __, 16, 18

14. $743 - \bigcirc = 21$

15. Estimate the difference: $508 - 216$. _____

16. Marie made 12 hot pads to sell. Her friend, Joanie, made 11 hot pads to sell with Marie. How many hot pads do the two girls have in all? _____

17. There are 342 girls and 357 boys at Mountain Elementary. How many girls and boys are there all together? _____

18. Of the 745 students at Highland Elementary, 592 either take the bus or get a ride to school. The rest of the students walk. How many students walk to school at Highland Elementary? _____

19. There are 847 stars in one section of the sky, as seen by a telescope at midnight. At 4:00 in the morning, only 291 stars can be seen by the same telescope. How many fewer stars are seen at 4:00 in the morning? _____

20. Mr. Davidson bought a new car in the years 1996, 2000, 2004, and 2008. If this pattern continues, in what years will Mr. Davidson buy his next 3 cars?

    _____

21. Ms. Baker had 18 desks in her room on Tuesday. On Wednesday, she had 7 left. How many desks were removed from her classroom?

    _____

22. Mrs. Farmer planted 13 roses in her garden on Monday and 17 more on Friday. About how many roses did she plant?

    _____

The problems below include both addition and subtraction in the same problem. Show your work by writing number sentences. (DOK 3)

23. Pine Valley Elementary school started the year with 823 students. During the year, 27 students moved away, and 12 new students enrolled. How many students were there at the end of the school year?

    _____

24. Ms. Wilson has 16 cups of flour in her cupboard. She uses 6 cups of flour to make bread. She then uses 2 more cups to make cookies. How many cups of flour does Ms. Wilson have left?

    _____

25. Jim collected 480 basketball cards. He gave Jessica 117 of them. Estimate the cards that Jim has left.

    _____

For additional practice, please see Chapter 2 Test located in the Teacher Guide.

Chapter 3
Multiplication

This chapter covers the following Grade 3 standards:

	Content Standard
Operations and Algebraic Thinking	3.OA.1, 3.OA.3, 3.OA.4, 3.OA.5, 3.OA.7, 3.OA.8
Number and Operations in Base 10	3.NBT.3

3.1 Using Repeated Addition (DOK 1)

Multiplication combines equal groups into one large group. It is a faster way to add the same number many times.

Example 1: Mr. Murphy is giving 2 muffins to each of his children. He has 4 children. How many muffins will he need?

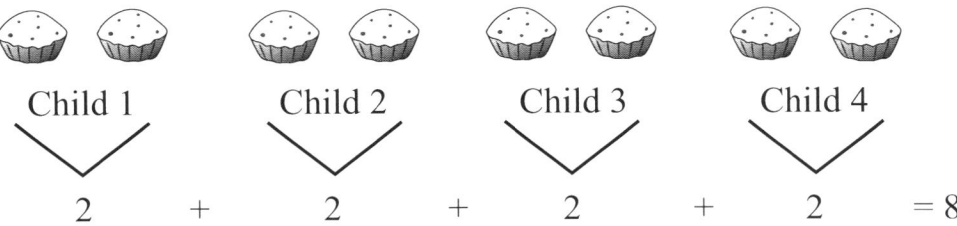

You can solve this problem by adding $2 + 2 + 2 + 2$. This is called **repeated addition**. Repeated addition is adding the same number many times. It is another way to write a multiplication problem. Another way to solve this would be to create a **multiplication sentence**.

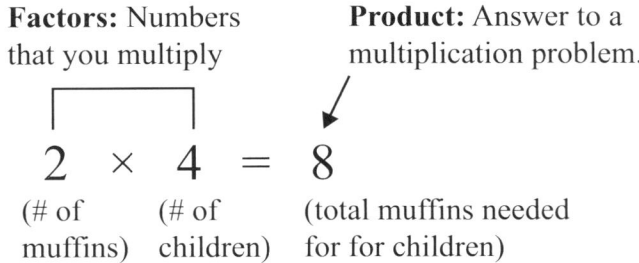

Factors: Numbers that you multiply

Product: Answer to a multiplication problem.

$$2 \times 4 = 8$$

(# of muffins) (# of children) (total muffins needed for for children)

Use what you know about repeated addition and multiplication to fill in the blank spaces in the table. (DOK 1)

	Repeated Addition	**Multiplication Sentence**	**Answer**
1.	3 + 3 + 3		
2.		4×3	
3.	2 + 2 + 2 + 2		
4.		5×2	
5.	3 + 3 + 3 + 3 + 3		
6.		7×2	
7.	1 + 1 + 1		
8.		6×3	
9.	8 + 8		
10.		2×3	

3.2 Using Arrays (DOK 1, 2)

Multiplication problems can also be solved by using arrays. An **array** is an arrangement of objects in rows and columns. The rows and columns form a rectangle. This rectangle is used as a tool to help understand multiplication.

Example 1: Use the array to write a multiplication sentence.

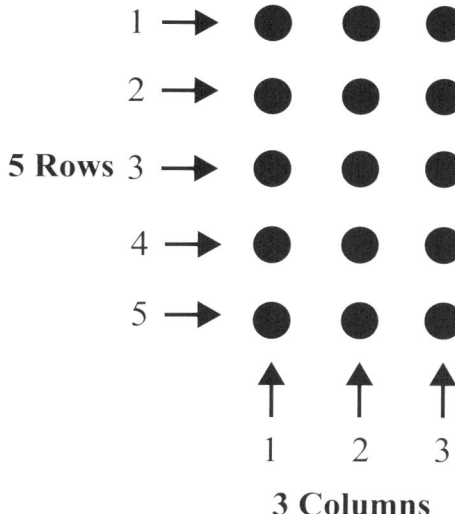

Step 1: Multiply the number of rows by the number of columns. This looks like 5×3.

Step 2: Find the answer by counting the number of circles in the array. There are 15 circles. This means that 5×3 is 15.

Example 2: Create an array to solve 3×7.

Step 1: Draw an array with 3 rows and 7 columns or 3 columns and 7 rows. Remember, columns go up and down. Rows go left to right.

● ● ● ● ● ● ●
● ● ● ● ● ● ●
● ● ● ● ● ● ●

Step 2: Count the number of dots.

Answer: Since there are 21 dots, $3 \times 7 = 21$.

Create an array, write a repeated addition sentence, and write a multiplication sentence. (DOK 3)

1. **6 rows, 2 columns**

   Array:

   Repeated Addition Sentence: _____
   Multiplication Sentence: _____

2. **3 rows, 4 columns**

   Array:

   Repeated Addition Sentence: _____
   Multiplication Sentence: _____

3. **4 rows, 6 columns**

   Array:

   Repeated Addition Sentence: _____
   Multiplication Sentence: _____

4. **2 rows, 2 columns**

   Array:

   Repeated Addition Sentence: _____
   Multiplication Sentence: _____

5. **5 rows, 4 columns**

   Array:

   Repeated Addition Sentence: _____
   Multiplication Sentence: _____

6. **3 rows, 2 columns**

   Array:

   Repeated Addition Sentence: _____
   Multiplication Sentence: _____

Use the array to write a multiplication sentence and to find the answer. (DOK 2)

7. ● ●
   ● ●

   Multiplication sentence: _____
   Answer: _____

8. ● ● ● ●
   ● ● ● ●

   Multiplication sentence: _____
   Answer: _____

9. ● ●
   ● ●
   ● ●

   Multiplication sentence: _____
   Answer: _____

10.
●
●
●
●

Multiplication sentence: _____

Answer: _____

11.
● ● ● ●
● ● ● ●
● ● ● ●
● ● ● ●

Multiplication sentence: _____

Answer: _____

12.
● ● ● ● ●
● ● ● ● ●
● ● ● ● ●

Multiplication sentence: _____

Answer: _____

13.
● ●
● ●
● ●
● ●

Multiplication sentence: _____

Answer: _____

14.
● ● ●
● ● ●
● ● ●

Multiplication sentence: _____

Answer: _____

3.3 Multiplication Properties (DOK 1, 2)

The **properties** (rules) of **multiplication** include the **identity property**, **associative property**, **commutative property**, and **distributive property**.

Property	Example	Rule
Identity	$7 \times 1 = 7$ $7 = 7$	Any number multiplied by one is equal to itself.
Commutative	$3 \times 4 = 4 \times 3$ $12 = 12$	When multiplying two numbers, the product will be the same no matter which order the two numbers are multiplied.
Associative	$(2 \times 4) \times 5 = 2 \times (4 \times 5)$ $8 \times 5 = 2 \times 20$ $40 = 40$	When multiplying three numbers, the product will be the same no matter which order the three numbers are multiplied.
Distributive	$5 \times (2 + 7) = (5 \times 2) + (5 \times 7)$ $5 \times 9 = 10 + 35$ $45 = 45$	When multiplying a number by the sum of two other numbers, the answer will be the same as multiplying the number by each of the two numbers in the parentheses and adding the results.

Identity Property of Multiplication

Rule: When you multiply a number by one, the number multiplied by one is equal to itself.

Examples: $8 \times 1 = 8$ $54 \times 1 = 54$ $367 \times 1 = 367$ $4 \times 1 = 4$

Fill in the blanks below. (DOK 1)

1. $16 \times 1 = \square$ 2. $\square \times 1 = 12$ 3. $28 \times \square = 28$

Commutative Property of Multiplication

Rule: It doesn't matter which order you multiply numbers, the answer is the same.

Examples: $9 \times 8 = 8 \times 9,$ $11 \times 10 = 10 \times 11$ $5 \times 6 = 6 \times 5$

Fill in the blanks below. (DOK 1)

4. $5 \times 7 = \boxed{} \times 5$
6. $4 \times 3 = \boxed{} \times 4$
8. $\boxed{} \times 5 = 5 \times 7$

5. $6 \times 2 = 2 \times \boxed{}$
7. $3 \times \boxed{} = 8 \times 3$
9. $9 \times 2 = 2 \times \boxed{}$

Associative Property of Multiplication

Rule: It doesn't matter how you group the numbers, the answer is the same.

Examples: $(5 \times 4) \times 3 = 5 \times (4 \times 3)$ $(6 \times 8) \times 2 = 6 \times (8 \times 2)$

Fill in the blanks below. (DOK 1)

10. $(5 \times 2) \times 4 = 5 \times (2 \times \boxed{})$
13. $(4 \times 6) \times 11 = 4 \times (6 \times \boxed{})$

11. $(6 \times 8) \times 2 = \boxed{} \times (8 \times 2)$
14. $(1 \times 9) \times 4 = 1 \times (\boxed{} \times 4)$

12. $(3 \times 7) \times 8 = \boxed{} \times (7 \times 8)$
15. $(3 \times \boxed{}) \times 2 = 3 \times (7 \times 2)$

The Distributive Property of Multiplication

The distributive property takes a problem like: $3 \times (2 + 4)$
and changes it to $(3 \times 2) + (3 \times 4)$.

In the first problem, you multiply 3 by the sum of $2 + 4$.
In the second problem, you multiply 3 by each of the numbers 2 and 4.
The problems look different, but they both have the same answer: 18.

$$3 \times (2 + 4) = 3 \times 6 = 18$$
$$(3 \times 2) + (3 \times 4) = 6 + 12 = 18$$

Remember, the distributive property allows the original problem, $3 \times (2 + 4)$, to change to $(3 \times 2) + (3 \times 4)$.

Fill in the blanks below. (DOK 1)

16. $7 \times (8 + 4) = (7 \times 8) + (7 \times \boxed{})$
19. $6 \times (7 + \boxed{}) = (6 \times 7) + (6 \times 2)$

17. $2 \times (6 + 3) = (\boxed{} \times 6) + (2 \times 3)$
20. $5 \times (5 + 8) = (5 \times \boxed{}) + (5 \times 8)$

18. $9 \times (1 + 4) = (9 \times 1) + (\boxed{} \times 4)$
21. $\boxed{} \times (3 + 9) = (8 \times 3) + (8 \times 9)$

3.4 Multiplication Facts (DOK 2)

Multiples can be found by multiplying a single number by 1, 2, 3, 4, and so on. The tables below show the first 3 multiples of the given numbers.

Given Number	First 3 Multiples
0	$0 \times 1 = \mathbf{0}, 0 \times 2 = \mathbf{0}, 0 \times 3 = \mathbf{0}$
1	$1 \times 1 = \mathbf{1}, 1 \times 2 = \mathbf{2}, 1 \times 3 = \mathbf{3}$
2	$2 \times 1 = \mathbf{2}, 2 \times 2 = \mathbf{4}, 2 \times 3 = \mathbf{6}$
3	$3 \times 1 = \mathbf{3}, 3 \times 2 = \mathbf{6}, 3 \times 3 = \mathbf{9}$
4	$4 \times 1 = \mathbf{4}, 4 \times 2 = \mathbf{8}, 4 \times 3 = \mathbf{12}$
5	$5 \times 1 = \mathbf{5}, 5 \times 2 = \mathbf{10}, 5 \times 3 = \mathbf{15}$

Special Cases to Remember:

Multiplying by 0: Any number multiplied by 0 is 0.

   Examples: $1 \times 0 = 0$ $5 \times 0 = 0$ $100 \times 0 = 0$

Multiplying by 1: Any number multiplied by 1 is itself.

   Examples: $1 \times 1 = 1$ $5 \times 1 = 5$ $100 \times 1 = 100$

Multiplying by 2: Any number multiplied by 2 is double itself and is always even.

   Examples: $1 \times 2 = 2$ $5 \times 2 = 10$ $100 \times 2 = 200$

Multiplying by 5: Any number multiplied by 5 ends in 0 or 5.

   Examples: $1 \times 5 = 5$ $5 \times 5 = 25$ $10 \times 5 = 50$

Use the multiplication facts to answer each question below (DOK 2).

1. $0 \times 3 =$ _____
2. $1 \times 7 =$ _____
3. $4 \times 2 =$ _____
4. $3 \times 5 =$ _____
5. $6 \times 0 =$ _____
6. $1 \times 15 =$ _____
7. $10 \times 2 =$ _____
8. $36 \times 0 =$ _____
9. $1 \times 90 =$ _____
10. $2 \times 2 =$ _____
11. $5 \times 4 =$ _____
12. $6 \times 5 =$ _____
13. $63 \times 0 =$ _____
14. $17 \times 1 =$ _____
15. $2 \times 6 =$ _____
16. $5 \times 7 =$ _____
17. $0 \times 28 =$ _____
18. $1 \times 66 =$ _____
19. $20 \times 2 =$ _____
20. $8 \times 5 =$ _____

List the first 5 multiples of each number below (DOK 1).

21. 6: _____

22. 7: _____

23. 8: _____

24. 9: _____

25. 10: _____

26. 11: _____

3.5 Multiplication Table (DOK 2)

A **multiplication table** can help develop multiplication skills. The numbers along the top and side are **factors**. The numbers in the middle of the table are the **products**.

Example 1: What is the product of 2×3?

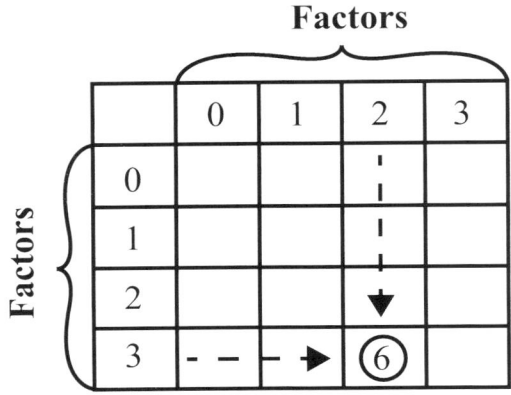

Step 1: Find row 3 and column 2. Then find the square where the row and column meet. Write the product of 2×3 in the square. Since the product is 6, write 6 in the square.

Fill in the following multiplication table. Some of the squares have been filled in for you. (DOK 2)

×	1	2	3	4	5	6	7	8	9	10
1				4						
2										20
3							21			
4			12		20				36	
5		10				30				
6										
7	7							56		
8										
9										
10										

Use the completed multiplication table as a tool to answer the questions below. (DOK 1)

1. $8 \times 9 =$ _____

2. $6 \times 4 =$ _____

3. $7 \times 5 =$ _____

4. $3 \times 6 =$ _____

5. $10 \times 10 =$ _____

6. $4 \times 9 =$ _____

7. $5 \times 5 =$ _____

8. $9 \times 9 =$ _____

9. $7 \times 9 =$ _____

10. $3 \times 10 =$ _____

11. $7 \times 7 =$ _____

12. $6 \times 2 =$ _____

13. $5 \times 4 =$ _____

14. $8 \times 3 =$ _____

15. $7 \times 8 =$ _____

16. $3 \times 2 =$ _____

17. $10 \times 8 =$ _____

18. $5 \times 9 =$ _____

19. $5 \times 6 =$ _____

20. $6 \times 6 =$ _____

21. $8 \times 4 =$ _____

22. $6 \times 9 =$ _____

23. $2 \times 1 =$ _____

24. $4 \times 7 =$ _____

25. $3 \times 3 =$ _____

26. $9 \times 2 =$ _____

27. $1 \times 7 =$ _____

28. $4 \times 6 =$ _____

29. $5 \times 2 =$ _____

30. $8 \times 6 =$ _____

31. $4 \times 2 =$ _____

32. $7 \times 3 =$ _____

33. $2 \times 2 =$ _____

34. $1 \times 10 =$ _____

35. $3 \times 9 =$ _____

36. $2 \times 7 =$ _____

37. $5 \times 8 =$ _____

38. $6 \times 3 =$ _____

39. $8 \times 8 =$ _____

3.6 Multiples of Ten (DOK 1, 2)

To find multiples of ten, multiply 10 by 1, 2, 3, 4, 5, and so on. The products 10, 20, 30, 40, 50, and so on are multiples of 10.
You can use multiples of 10 to simplify when multiplying.

Example 1: Find the product of 3 and 90.

$3 \times 90 = ?$

Multiple of 10

Step 1: Locate the multiple of 10. Ninety is the multiple of 10 because it ends in a zero.

Step 2: Remove the 0 from 90.

Step 3: Multiply.
$3 \times 9 = 27$

Step 4: Put the 0 back at the end of the answer: 270.

Find the product. (DOK 1)

1. $6 \times 80 =$

2. $4 \times 20 =$

3. $30 \times 7 =$

4. $50 \times 5 =$

5. $9 \times 60 =$

6. $8 \times 40 =$

7. $70 \times 3 =$

8. $60 \times 3 =$

9. $20 \times 6 =$

10. $8 \times 20 =$

11. $4 \times 50 =$

12. $9 \times 10 =$

13. $50 \times 7 =$

14. $6 \times 60 =$

15. $7 \times 60 =$

16. $8 \times 80 =$

17. $9 \times 50 =$

18. $10 \times 4 =$

19. $20 \times 7 =$

20. $9 \times 40 =$

3.7 Number Sentences (DOK 2)

Number Sentences can be used to show relationships between numbers. An **open number sentence** is a number sentence with one or more missing numbers.

Example 1: Find the missing value in the number sentence using multiplication facts.

$3 \times \boxed{} = 3$

Step 1: Decide which multiplication fact is being used.

$3 \times \boxed{} = 3$

Since the product is the same as the first factor, we know the missing number is 1.

Answer: The number sentence should read $3 \times 1 = 3$.

Example 2: Find the missing value in the number sentence using multiplication facts.

$5 \times \boxed{} = 0$

Step 1: Decide which multiplication fact is being used.

$5 \times \boxed{} = 0$

Since the product is 0, we know the missing number is 0.

Answer: The number sentence should read $5 \times 0 = 0$.

Example 3: Find the missing value in the number sentence using multiplication facts.

$\boxed{} \times 4 = 8$

Step 1: Decide which multiplication fact is being used.

$\boxed{} \times 4 = 8$

Since the product is double the first factor, we know the missing number is 2.

Answer: The number sentence should read $2 \times 4 = 8$.

Find the missing value. Show your work. (DOK 2)

21. $17 \times \square = 0$

26. $\square \times 12 = 12$

31. $18 \times \square = 18$

22. $1 \times \square = 24$

27. $10 \times \square = 20$

32. $\square \times 1 = 9$

23. $\square \times 8 = 16$

28. $\square \times 14 = 0$

33. $8 \times 10 = \square$

24. $\square \times 1 = 7$

29. $74 \times \square = 74$

34. $\square \times 56 = 0$

25. $23 \times \square = 0$

30. $\square \times 2 = 4$

35. $6 \times \square = 12$

3.8 Using Two Operations (DOK 3)

Parentheses in math are a grouping tool. We use parentheses to decide what parts should be looked at first. When given a math problem with parentheses, first simplify what is inside the parentheses. After simplifying inside the parentheses, simplify the rest of the problem.

Example 1: Simplify $(3 + 5) + (2 \times 5)$.

Step 1: Simplify the math inside each set of parentheses.
$(3 + 5) + (2 \times 5)$
$(8) + (10)$

Step 2: Simplify the rest of it.
$(8) + (10) = 18$

Answer: 18

Example 2: Simplify $(7 - 4) \times 9$.

    Step 1: Simplify the math inside the parentheses.
                $(7 - 4) \times 9$
                $(3) \times 9$

    Step 2: Simplify the rest of it
                $3 \times 9 = 27$

    Answer: 27

Simplify the problems below. (DOK 2)

1. $(8 \times 7) - 5 =$ _____

2. $6 + (12 \div 4) =$ _____

3. $(14 \times 2) + 7 =$ _____

4. $(3 \times 3) + (4 \times 3) =$ _____

5. $24 \times (6 - 4) =$ _____

6. $50 - (5 \times 3) =$ _____

7. $(7 \times 3) + 10 =$ _____

8. $(2 \times 4) + (2 \times 9) =$ _____

9. $(6 \times 7) + 11 =$ _____

10. $99 - (8 \times 3) =$ _____

11. $(4 \times 4) + 22 =$ _____

12. $(1 \times 9) + (6 \times 9) =$ _____

3.9 Multiplication Enrichment

1. Mr. Crowley arranged his arrowhead collection into 5 rows. Each row had 10 arrowheads. How many arrowheads were there in all? _____

2. Miss Kim has 6 packages of markers to share with her class. There are 8 markers in each package. How many markers does she have in all? _____

3. Each barrel of pickles at the state fair holds 80 pickles. There are 6 full barrels of pickles at the start of the day. How many pickles are there in all? _____

4. Sergio and his sister, Marissa, each saw 6 different deer while riding in the car. How many deer did they see in all? _____

5. Peter and his family went to a state park during summer vacation. Peter caught 3 fish each day. How many fish did Peter catch over 4 days? _____

6. Mrs. Jacobson made a bowl of punch for her family holiday party. Each family member had 2 cups of punch. There were 9 family members at the party. How many cups of punch did the family drink in all? _____

7. Mandy's pet fox had 4 babies each year in 2008, 2009, 2010, 2011, and 2012. How many babies did Mandy's pet fox have over the 5 years? _____

8. Steve and Henry were spinning a top and timing it to see how long it would spin before falling over. In the first 6 spins, the top spun exactly 8 seconds. How many seconds in all did the top spin? _____

9. There are 5 shelves on a book case. Luke puts 5 books on each of the first 3 shelves and 2 books on each of the next 2 shelves. How many books does Luke have? _____

10. There are 16 apples in each bag at the grocery store. There are 20 bags for sale. How many apples are there? _____

11. A package of 22 colored pencils costs $4. Sara spent $12 on colored pencils for her brothers and herself. How many colored pencils did Sara buy in all?

_____

12. Mrs. Tanner is making school lunches for her 5 children. She packs 4 carrot sticks, a sandwich, and a pear in each lunch. How many carrot sticks has Mrs. Tanner packed? _____

13. Rick sees 4 boxes of cereal in the cupboard. Each box holds 10 servings. How many servings are there in all 4 boxes? _____

Chapter 3 Review

Find the products. (DOK 2)

1. $7 \times 7 =$ _____

2. $8 \times 9 =$ _____

3. $6 \times 7 =$ _____

4. $5 \times 3 =$ _____

5. $4 \times 8 =$ _____

6. $9 \times 2 =$ _____

Fill in the small multiplication table below. (DOK 2)

7.

	3	4	5
3	9		
4			
5			25

Look at the arrays below. Write the multiplication sentences and simplify. (DOK 2)

8.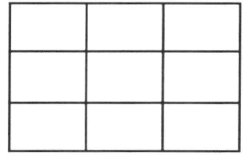

9.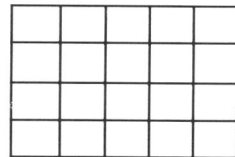

10.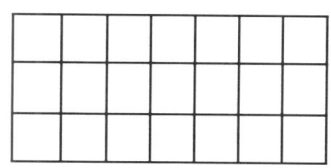

_____   _____   _____

Fill in the charts below by filling in the blanks. (DOK 1)

11. Multiples of 9s.

9	18			45	

12. Multiples of 4s.

4	8			20

13. Multiples of 8s.

8	16			

14. 2 + 2 + 2 = 6 _____   17. 8 + 8 = 16 _____

15. 6 + 6 + 6 + 6 = 24 _____   18. 7 + 7 + 7 = 21 _____

16. 3 + 3 + 3 + 3 = 12 _____   19. 5 + 5 + 5 = 15 _____

20. **Fill in the multiples of ten below. (DOK 1)**

$1 \times 10 =$ _____

$2 \times 10 =$ _____

$3 \times 10 =$ _____

$4 \times 10 =$ _____

$5 \times 10 =$ _____

21. **Write the answer on the spaces below. (DOK 2)**

$2 + (6 \times 10) =$ ———

$(4 + 5) \times (5 + 5) =$ ———

$(6 + 4) \times (5 + 2) =$ ———

$10 \times (2 + 6) =$ ———

$17 + (3 \times 10) =$ ———

Read each word problem and solve. Show your work. (DOK 2)

22. Amy bought 3 bags of balloons. There are 40 balloons in each bag. How many balloons did Amy buy?

_____

23. Which expression is equal to 12×7?

A) $(10 + 7) \times (2 + 7)$

B) $2 + (10 \times 7)$

C) $(6 \times 4) + (6 \times 3)$

D) $(6 + 7) \times (6 + 7)$

E) $(6 \times 7) + (6 \times 7)$

Read each word problem and solve. Show your work. (DOK 3)

24. There are 22 boxes of books. Four are empty. The other 18 boxes have 10 books each. How many total books are there?

_____

25. At the store, a package of 5 pens costs $1.00. Tom spent $8.00 on pens. How many pens did Tom buy?

_____

26. Find the missing value. (DOK 2)

$5 \times \boxed{} = 35$

27. Find the missing value. (DOK 2)

$4 \times 6 = \boxed{}$

28. Gianni has 7 packs of gum. Each pack of gum has 6 pieces in it. How many pieces does he have in all? (DOK 2) _____

29. Find the missing value. (DOK 2)

$\boxed{} \times 5 = 40$

For additional practice, please see Chapter 3 Test located in the Teacher Guide.

Chapter 4
Division

This chapter covers the following Grade 3 standards:

	Content Standard
Operations and Algebraic Thinking	3.OA.2, 3.OA.3, 3.OA.4, 3.OA.6, 3.OA.7, 3.OA.8

4.1 Using Repeated Subtraction (DOK 1)

Division separates one larger group into smaller, equal groups. It is a faster way to subtract the same amount many times.

Division can be written 3 ways.

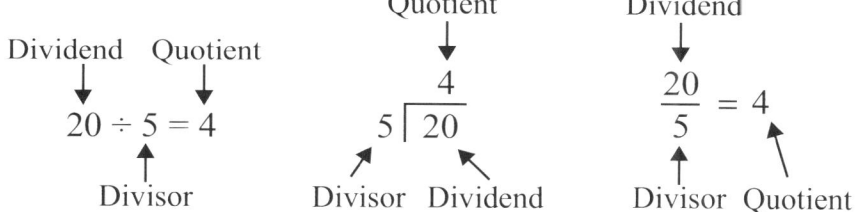

You can use repeated subtraction to divide. In **repeated subtraction**, the idea is to subtract the same amount until you reach zero. To divide 15 by 3, keep subtracting 3 from 15 until you get to 0. Look at the number line below.

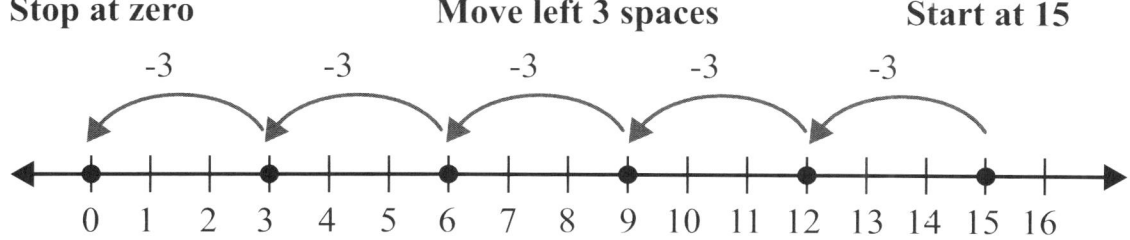

Count the number of times you moved left 3 spaces.

You moved left 3 spaces 5 times. This means that 15 divided by 3 is 5.

Example 1: Kaeley divides 25 playing cards among 5 friends. How many cards will each friend get?

Repeated Subtraction:

Circle the number of times you subtracted 5 to get to 0.

You subtracted 5 times. This means Kaeley will give each friend 5 cards!

Example 2: Joshua has 12 toy cars. He wants to give each of his friends 3 cars. How many friends will Joshua give his cars to? Use repeated subtraction to find the answer.

Step 1: List the information you know from the problem. Joshua has 12 toy cars.

Step 2: Write subtraction sentences. Keep subtracting until you reach 0.

$12 - 3 = 9$ $9 - 3 = 6$ $6 - 3 = 3$ $3 - 3 = 0$

Step 3: Count the number of times you subtracted.

1   2   3   4

$12 - 3 = 9$ $9 - 3 = 6$ $6 - 3 = 3$ $3 - 3 = 0$

There are 4 friends that will receive 3 cars each.
12 (cars in all) $\div$ 3 (cars) = 4 (friends)

Answer: $12 \div 3 = 4$. Joshua will give 4 friends 3 cars each.

Use repeated subtraction to find the missing factor. (DOK 1)

1. 16 ↗ 12 ↗ 8 ↗ 4 ↗ 0
    −4   −4   −4   −4

    $16 \div 4 = $ _____

2. 15 ↗ 12 ↗ 9 ↗ 6 ↗ 3 ↗ 0
    −3   −3   −3   −3   −3

    $15 \div 3 = $ _____

3. 21 ↗ 14 ↗ 7 ↗ 0
    −7   −7   −7

    $21 \div 7 = $ _____

4. 18 ↗ 15 ↗ 12 ↗ 9 ↗ 6 ↗ 3 ↗ 0
    −3   −3   −3   −3   −3   −3

    $18 \div 3 = $ _____

5. 18 ↗ 16 ↗ 14 ↗ 12 ↗ 10 ↗ 8 ↗ 6 ↗ 4 ↗ 2 ↗ 0
    −2   −2   −2   −2   −2   −2   −2   −2   −2

    $18 \div 2 = $ _____

6. 9 ↗ 0
    −9

    $9 \div 9 = $ _____

4.2 Using Arrays (DOK 1, 2)

Division problems can also be solved by using **arrays**. We can split the array into groups to see division.

Example 1: How many equal groups of 2 can be made with the array?

Step 1: Draw a circle around 2 blocks at a time.

Step 2: Count how many groups you circled.

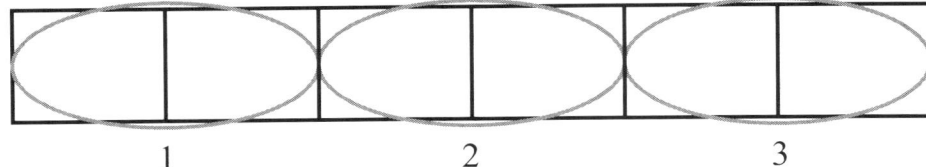

Step 3: So 6 divided by 2 is 3.

Example 2: Mrs. Jackson has 15 pencils. She wants to put them into 3 equal groups. How many pencils will be in each group? Write the number sentence.

Step 1: List the information you know from the problem.
Mrs. Jackson has 15 pencils.

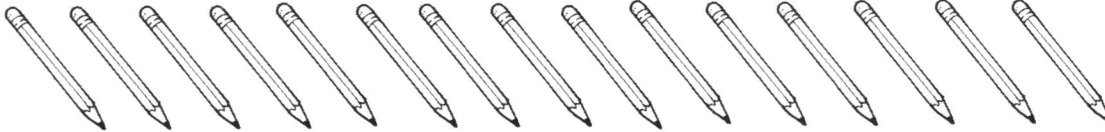

Step 2: Make three circles to represent the groups.

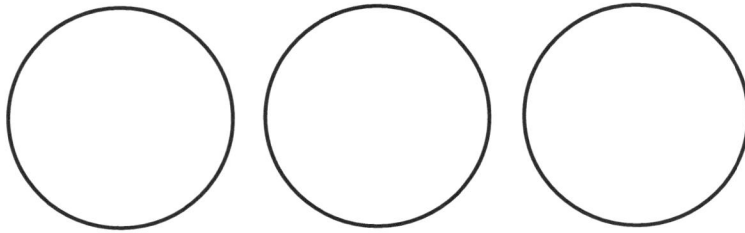

Step 3: Start placing a pencil into each circle until all the pencils are used.

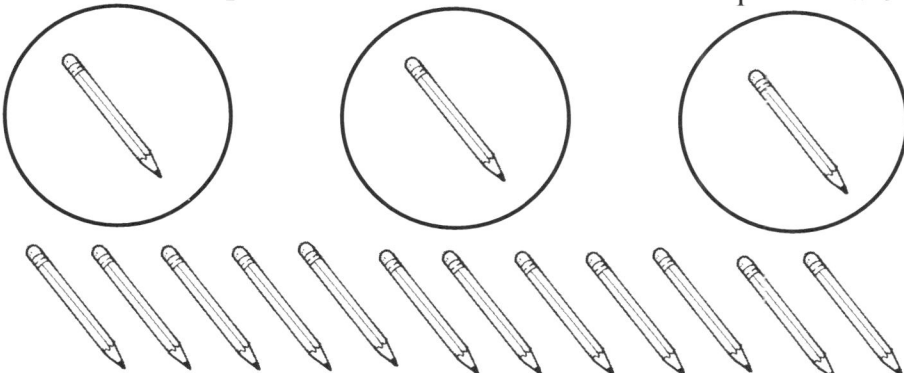

Step 4: Count how many pencils are equally divided into the 3 groups.

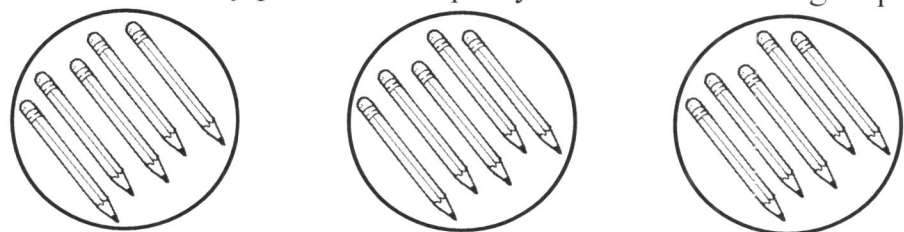

There are 5 pencils in each group.
15 (pencils in all) ÷ 3 (groups) = 5 (pencils)

Answer: 15 ÷ 3 = 5. Mrs. Jackson can place 5 pencils equally into the 3 groups.

Use the information to divide the items into equal groups. Write the number sentence on the line. (DOK 2)

1. How many equal groups of 2 can be made with the shells below? _____

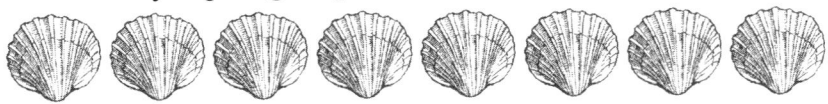

2. How many equal groups of 5 can be made with the sea horses below? _____

3. How many equal groups of 3 can be made with the sand dollars shown below? _____

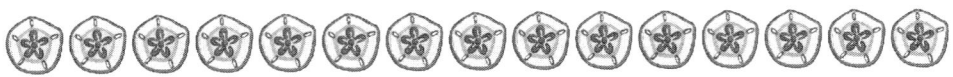

4. How many equal groups of 4 can be made with the pails below? _____

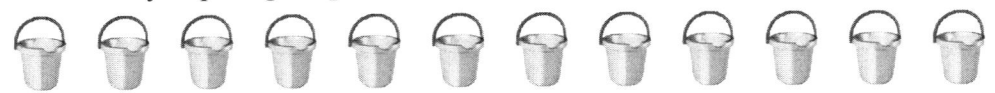

5. How many equal groups of 3 can be made with the starfish below? _____

6. How many equal groups of 2 can be made with the Teddy bears below? _____

7. How many equal groups of 3 can be made with the toy robots below? _____

For each of the division problems, find how many groups you can make. Write your answer in the blank. Show your work by circling each group. (DOK 2)

8. $10 \div 5 =$ _____ ★ ★ ★ ★ ★ ★ ★ ★ ★ ★ ★ ★ ★ ★ ★ ★

9. $15 \div 3 =$ _____ ☘ ☘ ☘ ☘ ☘ ☘ ☘ ☘ ☘ ☘ ☘ ☘ ☘ ☘ ☘

10. $16 \div 4 =$ _____ ■ ■ ■ ■ ■ ■ ■ ■ ■ ■ ■ ■ ■ ■

11. $18 \div 6 =$ _____ ◆ ◆ ◆ ◆ ◆ ◆ ◆ ◆ ◆ ◆ ◆ ◆ ◆

12. $15 \div 3 =$ _____ ✵ ✵ ✵ ✵ ✵ ✵ ✵ ✵ ✵ ✵ ✵ ✵ ✵

13. $16 \div 8 =$ _____ ◆ ◆ ◆ ◆ ◆ ◆ ◆ ◆ ◆ ◆ ◆ ◆ ◆ ◆ ◆ ◆

14. $18 \div 3 =$ _____ ⚾ ⚾ ⚾ ⚾ ⚾ ⚾ ⚾ ⚾ ⚾

15. $20 \div 4 =$ _____ ✿ ✿ ✿ ✿ ✿ ✿ ✿ ✿ ✿ ✿ ✿ ✿ ✿ ✿

16. $6 \div 2 = 3$ _____ A)

17. $12 \div 3 = 4$ _____ B)

18. $8 \div 4 = 2$ _____ C)

19. $9 \div 3 = 3$ _____ D)

4.3 Using Multiplication for Division (DOK 2)

Example 1: Solve: $21 \div 3$

Step 1: Find 3 on the multiplication table.

×	1	2	3	4	5	6	7	8	9	10
1	1	2	3	4	5	6	7	8	9	10
2	2	4	6	8	10	12	14	16	18	20
③	3	6	9	12	15	18	21	24	27	30
4	4	8	12	16	20	24	28	32	36	40
5	5	10	15	20	25	30	35	40	45	50
6	6	12	18	24	30	36	42	48	54	60
7	7	14	21	28	35	42	49	56	63	70
8	8	16	24	32	40	48	56	64	72	80
9	9	18	27	36	45	54	63	72	81	90
10	10	20	30	40	50	60	70	80	90	100

Step 2: Follow the row that the 3 is on until you find 21.

×	1	2	3	4	5	6	7	8	9	10
1	1	2	3	4	5	6	7	8	9	10
2	2	4	6	8	10	12	14	16	18	20
③	3	6	9	12	15	18	21	24	27	30
4	4	8	12	16	20	24	28	32	36	40
5	5	10	15	20	25	30	35	40	45	50
6	6	12	18	24	30	36	42	48	54	60
7	7	14	21	28	35	42	49	56	63	70
8	8	16	24	32	40	48	56	64	72	80
9	9	18	27	36	45	54	63	72	81	90
10	10	20	30	40	50	60	70	80	90	100

Step 3: Find the column that 21 falls in by following the column to the top of the chart.

×	1	2	3	4	5	6	7	8	9	10
1	1	2	3	4	5	6		8	9	10
2	2	4	6	8	10	12	14	16	18	20
3	3	6	9	12	15	18	21	24	27	30
4	4	8	12	16	20	24	28	32	36	40
5	5	10	15	20	25	30	35	40	45	50
6	6	12	18	24	30	36	42	48	54	60
7	7	14	21	28	35	42	49	56	63	70
8	8	16	24	32	40	48	56	64	72	80
9	9	18	27	36	45	54	63	72	81	90
10	10	20	30	40	50	60	70	80	90	100

Answer: $21 \div 3 = 7$

Use the multiplication table below to solve the division problems. (DOK 1)

×	1	2	3	4	5	6	7	8	9	10
1	1	2	3	4	5	6	7	8	9	10
2	2	4	6	8	10	12	14	16	18	20
3	3	6	9	12	15	18	21	24	27	30
4	4	8	12	16	20	24	28	32	36	40
5	5	10	15	20	25	30	35	40	45	50
6	6	12	18	24	30	36	42	48	54	60
7	7	14	21	28	35	42	49	56	63	70
8	8	16	24	32	40	48	56	64	72	80
9	9	18	27	36	45	54	63	72	81	90
10	10	20	30	40	50	60	70	80	90	100

1. $35 \div 5 =$ _____

2. $81 \div 9 =$ _____

3. $56 \div 8 =$ _____

4. $36 \div 6 =$ _____

5. $14 \div 7 =$ _____

6. $9 \div 1 =$ _____

7. $16 \div 2 =$ _____

8. $42 \div 6 =$ _____

9. $15 \div 5 =$ _____

10. $10 \div 10 =$ _____

11. $24 \div 4 =$ _____

12. $64 \div 8 =$ _____

13. $54 \div 9 =$ _____

14. $27 \div 3 =$ _____

15. $30 \div 5 =$ _____

16. $12 \div 2 =$ _____

17. $18 \div 3 =$ _____

18. $40 \div 10 =$ _____

19. $20 \div 2$ _____

20. $8 \div 4$ _____

21. $9 \div 3$ _____

22. $10 \div 5$ _____

23. $16 \div 4$ _____

24. $18 \div 9$ _____

25. $7\overline{)14} =$ _____

26. $7\overline{)63} =$ _____

27. $2\overline{)20} =$ _____

28. $2\overline{)16} =$ _____

29. $8\overline{)56} =$ _____

30. $6\overline{)48} =$ _____

31. $\dfrac{81}{9} =$ _____

32. $\dfrac{36}{6} =$ _____

33. $\dfrac{20}{4} =$ _____

34. $\dfrac{21}{3} =$ _____

35. $\dfrac{64}{8} =$ _____

36. $\dfrac{28}{4} =$ _____

4.4 Division Facts (DOK 2)

Dividing with 0: Zero divided by any number is 0. However, you cannot divide a number by 0.

Example 1: Dividing 0 cookies into 5 groups will give you 0 cookies.

Example 2: Dividing 5 cookies into 0 groups is impossible.

Dividing with 1: A number divided by 1 is the same number.

Example: $17 \div 1 = 17$

Dividing with 2: All even numbers are divisible by 2.

Examples: $8 \div 2 = 4$ $\qquad$ $34 \div 2 = 17$ $\qquad$ $48 \div 2 = 24$

Dividing with 3: If digits of a number add to a multiple of 3, then that number is divisible by 3.

Example: $12 \Rightarrow 1 + 2 = 3$, so 12 is divisible by 3

Dividing with 5: All numbers ending in 0 or 5 are divisible by 5.

Examples: $10 \div 5 = 2$ $\qquad$ $25 \div 5 = 5$ $\qquad$ $90 \div 5 = 18$

Dividing with 6: Any number is divisible by 6 if it is also divisible by 2 and 3.

Example: $12 \div 2 = 6$ and $12 \div 3 = 4$, so 12 is divisible by 6 $\Rightarrow 12 \div 6 = 2$

Dividing with 9: If digits of a number add to a multiple of 9, then that number is divisible by 9.

Example: $54 \Rightarrow 5 + 4 = 9$, so 54 is divisible by 9

Dividing with 10: All numbers ending in 0 are divisible by 10.

Examples: $100 \div 10 = 10$ $\qquad$ $30 \div 10 = 3$ $\qquad$ $150 \div 10 = 15$

Circle Yes or No. If Yes, give the correct answer. (DOK 2)

1. Is 56 divisible by 10? Yes   No _____
   Which division fact did you use?_____

2. Is 32 divisible by 2? Yes   No _____
   Which division fact did you use?_____

3. Is 49 divisible by 5? Yes   No _____
   Which division fact did you use?_____

4. Is 99 divisible by 9? Yes   No _____
   Which division fact did you use?_____

5. Is 32 divisible by 1? Yes   No _____
   Which division fact did you use?_____

6. Is 94 divisible by 0? Yes   No _____
   Which division fact did you use?_____

7. Is 70 divisible by 10? Yes   No _____
   Which division fact did you use?_____

8. Is 63 divisible by 9? Yes   No _____
   Which division fact did you use?_____

9. Is 52 divisible by 2? Yes   No _____
   Which division fact did you use?_____

10. Is 85 divisible by 5? Yes   No _____
    Which division fact did you use?_____

11. Is 0 divisible by 21? Yes   No _____
    Which division fact did you use?_____

12. Is 56 divisible by 1? Yes   No _____
    Which division fact did you use?_____

4.5 Fact Families (DOK 2)

A **fact family** is a group of 3 numbers that are related through multiplication and division.

Example 1: Find the fact family for 2, 5, and 10.

$$2 \times 5 = 10$$
$$5 \times 2 = 10$$
$\Big\rangle$ Multiplication Number Sentences

$$10 \div 2 = 5$$
$$10 \div 5 = 2$$
$\Big\rangle$ Division Number Sentences

Multiplication and Division Fact Family Rules

Rule Number 1: The two smaller numbers multiplied together will equal the largest number.

Rule Number 2: The largest number divided by one of the smaller numbers will equal the other small number.

Find the fact families for the following numbers. (DOK 2)

1. 7, 5, 35

3. 14, 2, 7

5. 9, 8, 72

2. 1, 6, 6

4. 3, 6, 18

6. 54, 6, 9

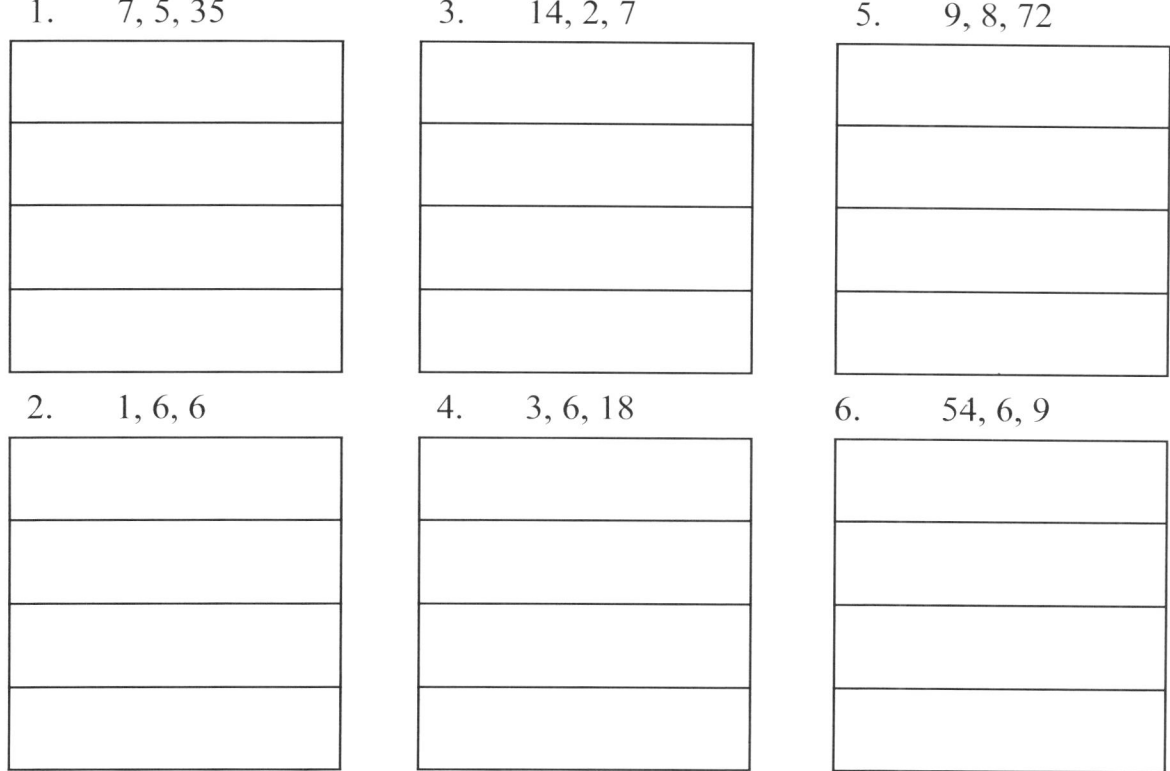

7.   2, 18, 9

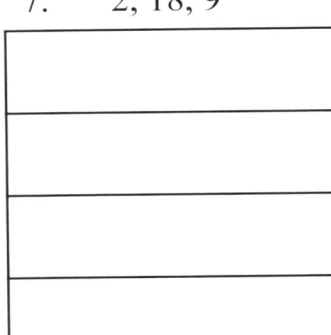

9.   42, 6, 7

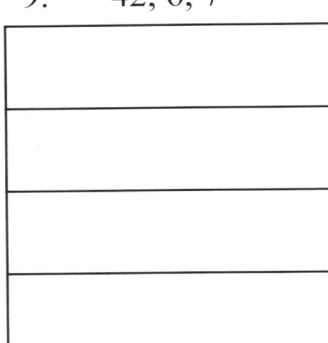

11.   7, 8, 56

8.   4, 16, 4

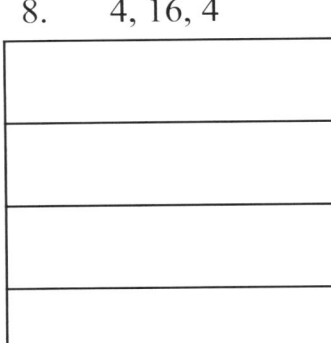

10.   5, 40, 8

12.   48, 8, 6

Circle the number sentence that does not belong in each fact family. (DOK 2)

13.   A) $6 \times 8 = 48$ B) $48 \div 6 = 8$ C) $48 \div 6 = 6$ D) $8 \times 6 = 48$

14.   A) $2 \times 7 = 14$ B) $7 \times 2 = 7$ C) $7 \times 2 = 14$ D) $14 \div 2 = 7$

15.   A) $36 \div 9 = 4$ B) $9 \times 4 = 36$ C) $4 \times 9 = 39$ D) $36 \div 4 = 9$

16.   A) $28 \div 7 = 7$ B) $28 \div 7 = 4$ C) $4 \times 7 = 28$ D) $7 \times 4 = 28$

17.   A) $18 \div 3 = 4$ B) $18 \div 3 = 6$ C) $3 \times 6 = 18$ D) $6 \times 3 = 18$

4.6 Number Sentences (DOK 2)

Multiplication and division are opposite operations of each other. When solving for missing values in a division sentence, use fact families. When solving for missing values in a multiplication number sentence, use division.

Example 1: Find the missing value.

$27 \div \square = 3$

Step 1: In a fact family, we know that the larger number can be divided by either of the smaller numbers. Rewrite the problem to solve.

$27 \div \square = 3$

$27 \div 3 = \square$

Step 2: Divide: $27 \div 3 = 9$

Answer: 9

Example 2: Find the missing value.

$6 \times \square = 24$

Step 1: Since this is a multiplication sentence, use division to solve.

Step 2: Divide: $24 \div 6 = 4$

Answer: 4

Rules when finding missing values of number sentences

Rule 1: If the $\square$ is the first number in a division problem, multiply the two smaller numbers. $\square \div 3 = 4$

Multiply $3 \times 4 = 12$

12 is the missing number.

Rule 2: If the $\square$ is the last number in a multiplication sentence, multiply the two smaller numbers to solve.

$7 \times 2 = \square$

$7 \times 2 = 14$

14 is the missing number.

Write the missing value in the box. (DOK 2)

1. $63 \div \boxed{} = 9$

2. $48 \div \boxed{} = 6$

3. $25 \div \boxed{} = 5$

4. $\boxed{} \div 2 = 8$

5. $4 \div \boxed{} = 2$

6. $\boxed{} \div 10 = 4$

7. $72 \div \boxed{} = 8$

8. $\boxed{} \div 3 = 8$

9. $18 \div \boxed{} = 3$

10. $\boxed{} \div 2 = 10$

11. $\boxed{} \div 7 = 5$

12. $\boxed{} \div 4 = 9$

13. $42 \div \boxed{} = 7$

14. $\boxed{} \div 8 = 8$

15. $10 \div \boxed{} = 10$

16. $\boxed{} \div 3 = 6$

17. $54 \div \boxed{} = 9$

18. $12 \div \boxed{} = 6$

19. $\boxed{} \div 5 = 7$

20. $36 \div \boxed{} = 6$

21. $\boxed{} \div 1 = 7$

22. $15 \div \boxed{} = 3$

23. $27 \div \boxed{} = 9$

24. $\boxed{} \div 2 = 8$

25. $32 \div \boxed{} = 4$

26. $\boxed{} \div 3 = 5$

27. $36 \div \boxed{} = 4$

28. $20 \div \boxed{} = 4$

29. $\boxed{} \div 4 = 8$

30. $100 \div \boxed{} = 10$

31. $90 \div \boxed{} = 9$

32. $\boxed{} \div 4 = 2$

33. $10 \div \boxed{} = 5$

34. $9 \div \boxed{} = 3$

35. $\boxed{} \div 3 = 4$

36. $42 \div \boxed{} = 7$

37. $81 \div \boxed{} = 9$

38. $12 \div \boxed{} = 2$

4.7 Division Enrichment

1. John has 24 toy cars. He has an equal number of black, red, and blue cars. He divides the toy cars equally onto 2 shelves in his bedroom. There is an equal number of each color on each shelf. How many blue cars are on one shelf?

2. Abby has 40 hair ribbons. She has an equal number of plain ribbons, sparkle ribbons, and striped ribbons. She puts the hair ribbons into 2 plastic bags. Each bag has an equal number of ribbons of each kind. How many sparkle ribbons are in each bag?

3. Ms. Cole has 88 buttons. She has an equal number of white, red, gold, and blue buttons. She puts the buttons into 2 plastic bags. There are two colors in each bag. How many buttons are in each bag?

4. Jill has 2 cats and 1 dog. She wants them to share 9 ounces of cooked, ground chicken meat. How many ounces will each pet get if they each get an equal share?

5. There are 8 paint brushes for 4 students. How many paint brushes will each student get if they divide the paint brushes equally?

6. Grandma made a cake for Elian's birthday. There are 12 pieces of cake in all. If there are 6 people at the birthday party and each person gets an equal amount of cake, how many pieces of cake will each person get?

7. Tanya is collecting buttons to sew on a pillow for decoration. So far she has 26 buttons sewn on the pillow. Her grandmother gave her 6 more buttons. Tanya divides the total number of buttons she has into 4 groups. How many buttons are in each group?

8. A pet store divides 40 hamsters equally among 10 cages. Make a number sentence to show how many hamsters are in each cage.

Chapter 4 Review

Solve the division problems. (DOK 1)

1. $24 \div 8 =$ _____

2. $60 \div 6 =$ _____

3. $36 \div 9 =$ _____

4. $42 \div \boxed{} = 7$

5. $35 \div \boxed{} = 5$

6. $56 \div \boxed{} = 7$

7. Use repeated subtraction to find the quotient.

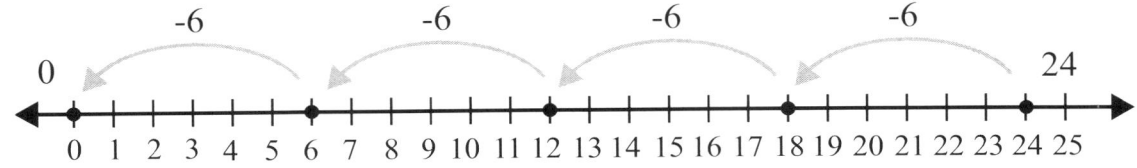

$24 \div 6 = \boxed{}$

8. Given the divided array, complete the division sentence.

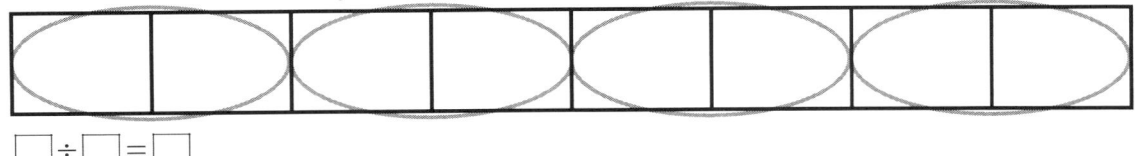

$\boxed{} \div \boxed{} = \boxed{}$

Read each problem carefully and solve. (DOK 2)

9. Sara shared her stuffed animal collection with her little sister. She had 14 stuffed animals. If Sara and her sister each got an equal share of the stuffed animals, how many did each girl get?
   $\boxed{}$ stuffed animals

10. How many groups of 2 can be made with the pennies below? _____

11. Is 330 divisible by 10?   Yes   No

12. Is 41 divisible by 2?   Yes   No

13. Is 25 divisible by 5?   Yes   No

14. Complete the number sentence.
    $\Box \div 7 = 3$

15. Complete the number sentence.
    $\Box \div 9 = 8$

16. Complete the number sentence.
    $\Box \div 10 = 5$

17. Mrs. Anderson made chocolate chip cookies for her kids. There are 15 cookies and 3 kids. If the cookies are divided evenly, how many cookies will each kid get?

    $\Box$ cookies

Write the fact family for each set of numbers. (DOK 2)

18.   8, 7, 56   19.   36, 4, 9   20.   2, 7, 14

21. How many groups of 3 can be made from the fish in the picture below?
(DOK 2) _____

**Use the multiplication table to the right to solve the division problems.
(DOK 2)**

22. $20 \div 4 =$ _____

23. $18 \div 6 =$ _____

24. $9 \div 3 =$ _____

25. $8 \div 2 =$ _____

26. $16 \div 4 =$ _____

×	0	1	2	3	4	5
0	0	0	0	0	0	0
1	0	1	2	3	4	5
2	0	2	4	6	8	10
3	0	3	6	9	12	15
4	0	4	8	12	16	20
5	0	5	10	15	20	25

For additional practice, please see Chapter 4 Test located in the Teacher Guide.

Chapter 5
Fractions

This chapter covers the following Grade 3 standards:

	Content Standard
Number and Operations - Fractions	3.NF.1, 3.NF.2, 3.NF.3, 3.MD.4

5.1 Fraction Facts (DOK 1)

A **fraction** is a number divided by another number. Fractions show how many portions of a whole there are. In a fraction, the number on top is called the **numerator**. The number on bottom is called the **denominator**.

Fraction: $\dfrac{\text{number of parts you have}}{\text{total number of parts}} = \dfrac{\text{numerator}}{\text{denominator}}$

Example 1: You cut an orange into 6 slices and eat 1 slice. What portion of the orange have you eaten?

Step 1: How many total parts do you have? Since you cut the orange into 6 slices, there are 6 total parts. So the denominator is 6.

Step 2: How many slices did you eat? You ate 1 slice. So the numerator is 1.

Step 3: Rewrite the information in fraction form.
$\dfrac{\text{number of parts you have}}{\text{total number of parts}} = \dfrac{\text{numerator}}{\text{denominator}} = \dfrac{1}{6}$

When reading a fraction out loud, $\dfrac{1}{4}$ is said "one-fourth" or "one over four." As you can see, you call the numerator by the number's name. To read the denominator, follow the chart below.

Denominator number	How to say it...
2	half
3	third
4	fourth
5	fifth
6	sixth

The pattern continues this way. If the numerator is greater than 1, simply add an "s" to the denominator unless the denominator is 2. For example, $\dfrac{2}{3}$ is said "two-thirds."

Write the fractions in word form. (DOK 1)

1. $\dfrac{2}{3}$ _____

2. $\dfrac{1}{2}$ _____

3. $\dfrac{4}{6}$ _____

4. $\dfrac{7}{8}$ _____

5. $\dfrac{1}{4}$ _____

6. $\dfrac{1}{3}$ _____

7. $\dfrac{2}{6}$ _____

8. $\dfrac{3}{8}$ _____

9. $\dfrac{1}{6}$ _____

10. $\dfrac{3}{4}$ _____

11. $\dfrac{5}{8}$ _____

12. $\dfrac{5}{6}$ _____

Write the fractions in standard form. (DOK 1)

13. two-sixths _____

14. one-eighth _____

15. three-fourths _____

16. four-eighths _____

17. seven-eighths _____

18. one-fourth _____

19. five-eighths _____

20. three-sixths _____

21. one-sixth _____

22. two-thirds _____

23. one-half _____

24. one-sixth _____

25. two-fourths _____

26. four-sixths _____

27. one-third _____

28. four-eighths _____

5.2 Fractions of a Whole (DOK 1, 2)

Fractions of a whole can be found by dividing one object into equal parts.

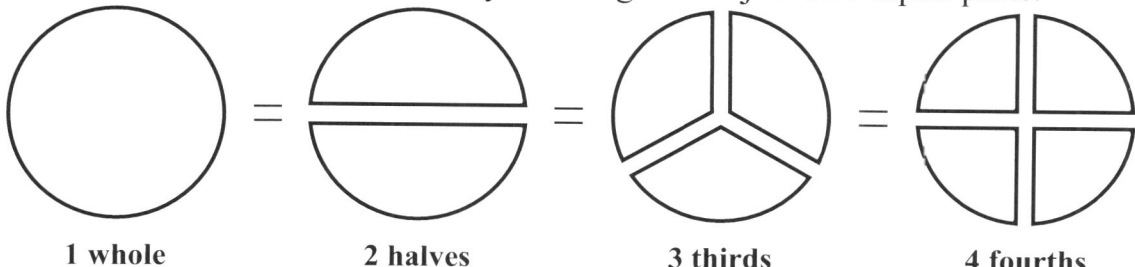

Two-halves is written as $\frac{2}{2}$. Since we still have all parts of the first image, $\frac{2}{2}$ is a whole, or 1. The same idea follows with $\frac{3}{3}$, $\frac{4}{4}$, and so on.

Example 1: Write the shaded part of the model as a fraction.

    Step 1: Look at the model. Count the total number of parts.

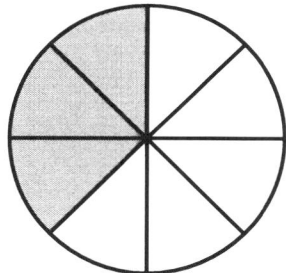

            There are 8 total parts.

    Step 2: Place the total parts in the denominator.

$$\frac{}{8}$$

    Step 3: Now count the amount of shaded parts.

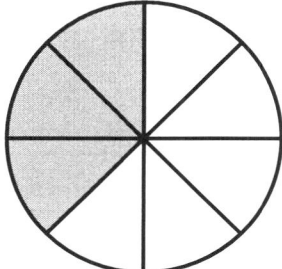

            There are 3 shaded parts.

Step 4: Place the number of shaded pieces in the numerator. $\frac{3}{8}$

Answer: $\frac{3}{8}$ of the model is shaded. This is read out loud as "three-eighths."

Write a fraction to name the shaded pieces of the model. (DOK 1)

1. 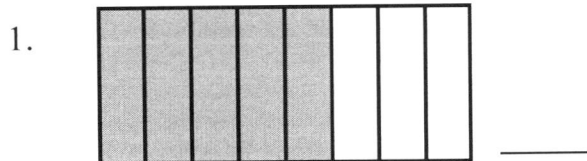 _____

2. 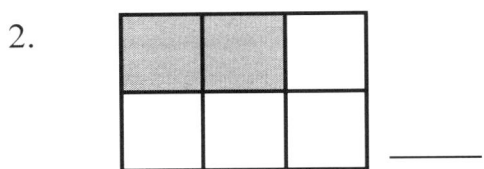 _____

3. 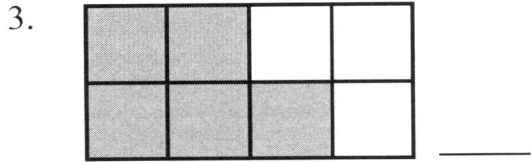 _____

4. 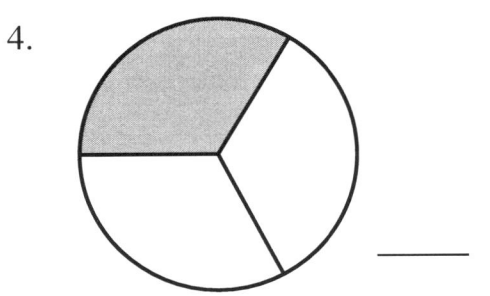 _____

5. 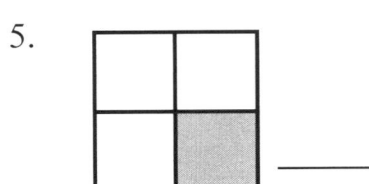 _____

6.

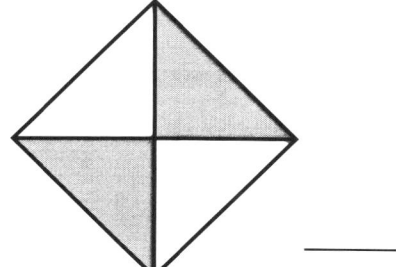 _____

7.

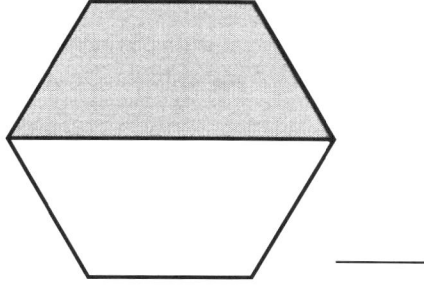 _____

8.

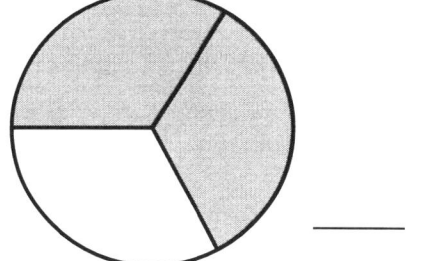 _____

9.

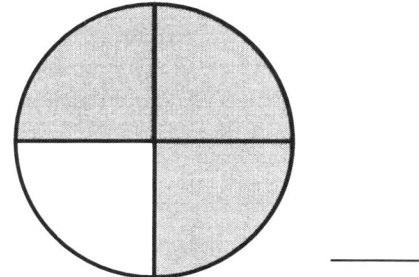 _____

10. Leticia and her 3 sisters will share a pizza equally. What fraction of the pizza will Leticia get?

11. Mrs. Jackson has 6 students in her music class. She will give each student an equal part of her homemade pie. What fraction of the pie will each student get?

12. Shauna divided her granola bar into 6 equal parts. She ate 5 parts. What fraction of the granola bar has Shauna eaten?

13. Leon and his 2 friends will evenly share a cupcake. What fraction of the cupcake will Leon get?

5.3 Fractions of a Group (DOK 2)

A fraction of a whole is when one item is divided or split into parts. A fraction of a group is when a number of things are separated into parts.

Example 1: Josie has 6 teddy bears. Three teddy bears are brown. The other three teddy bears are gray. What fraction of teddy bears are brown?

   Step 1: Josie has a total of 6 teddy bears. This is the denominator.

   Step 2: Three of the teddy bears are brown. This is the numerator.

   Step 3: Rewrite the information as a fraction. $\frac{3}{6}$

   Answer: $\frac{3}{6}$ or "three-sixths" of the teddy bears are brown.

Example 2: There are 7 ducks in a pond. All 7 of the ducks are swimming. What fraction of the group of ducks is swimming?

   Step 1: There is a total of 7 ducks swimming. This is the denominator.

   Step 2: We want to find the fraction of the group that is swimming. All 7 ducks are swimming. This is the numerator.

   Step 3: Rewrite the information as a fraction. $\frac{7}{7}$

   Answer: $\frac{7}{7}$ or "seven-sevenths" of the ducks are swimming. Since 7 out of 7 ducks are swimming, we would state that the whole group (1) is swimming.

Choose the model to fit the fraction problems below. (DOK 2)

1. Which model shows $\frac{1}{3}$ of the strawberries with a bite out of them? _____

2. Which model shows $\frac{4}{6}$ of the strawberries with a bite out of them? _____

3. Which model shows $\frac{1}{4}$ of the strawberries with a bite out of them? _____

4. Which model shows $\frac{3}{8}$ of the strawberries with a bite out of them? _____

5. Which model shows $\frac{1}{2}$ of the strawberries with a bite out of them? _____

6. Which model shows $\frac{7}{8}$ of the strawberries with a bite out of them? _____

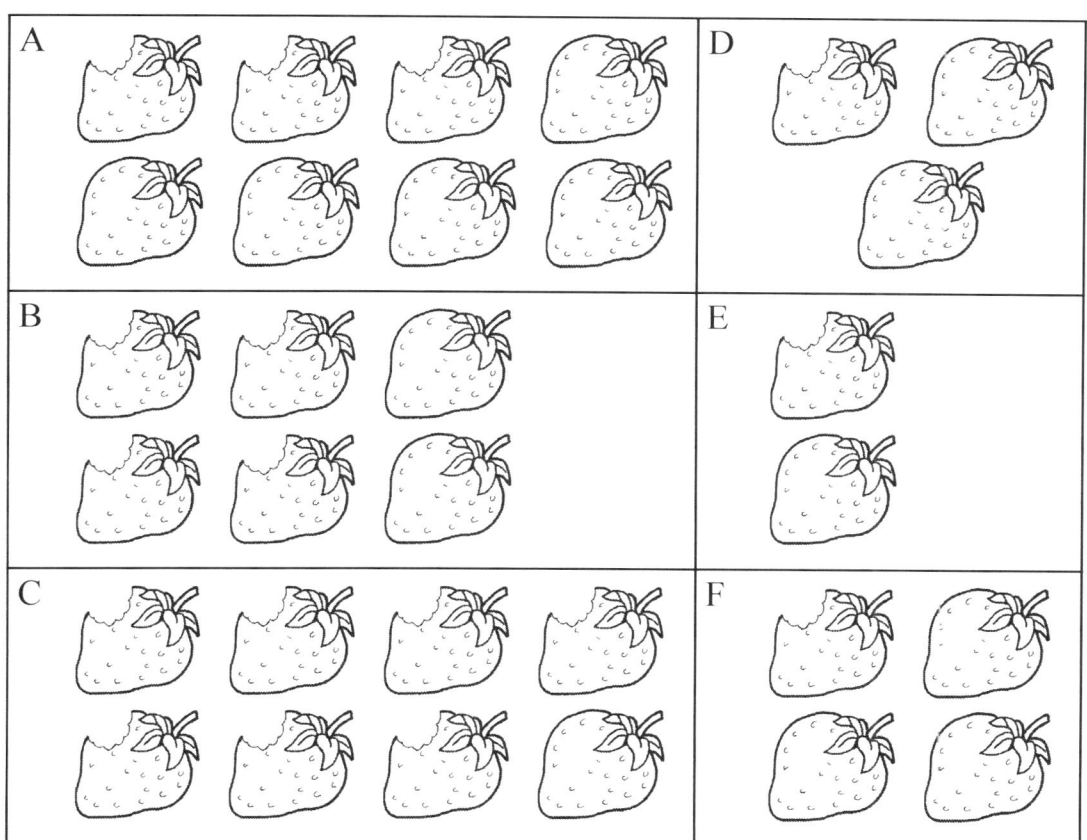

Match the problem on the left to the fraction model on the right. (DOK 3)

7. Mia has a bag of beads. One-fourth of the beads are red, one-half are blue, and one-fourth are white. Which model shows Mia's bag of beads divided into colors?

A)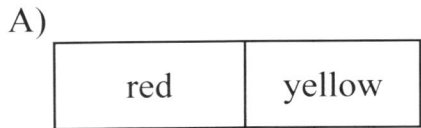

8. Kevin has one-half red toy cars, one-fourth blue toy cars, and one-fourth white toy cars. Which model shows how Kevin's toy cars are divided into colors?

B)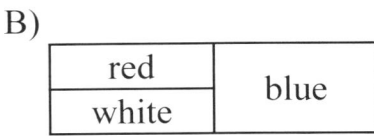

9. Sophia has one-half white tops, one-fourth red tops, and one-fourth blue tops. Which model shows how Sophia's tops are divided into colors?

C)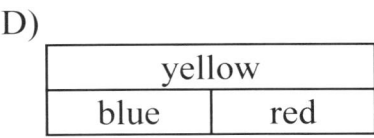

10. Harold has one-fourth yellow pencils, one-fourth red pencils, and one-half blue pencils. Which model shows how Harold's pencils are divided into colors?

D)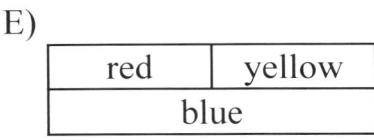

11. Kim has one-half yellow ribbons, one-fourth blue ribbons, and one-fourth red ribbons. Which model shows how Kim's ribbons are divided into colors?

E)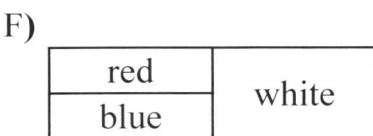

12. James has one-half yellow toy cars, one-fourth red toy cars, and one-fourth white toy cars. Which model shows how James' toy cars are divided into colors?

F)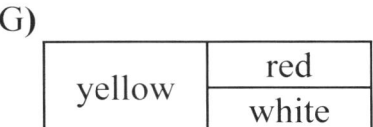

13. Lisa has one-half red beads, and one-half yellow beads. Which choice shows how Lisa's beads are divided into colors?

G)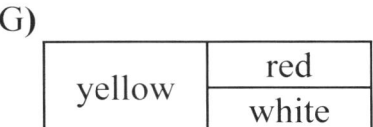

Use the information to write a fraction. (DOK 2)

14. Brittany has 6 barrettes. Four are heart barrettes and 2 are circle barrettes. Write a fraction to show how many are heart barrettes.

     ———————

15. There are 6 fruitsicles left in the freezer. Write a fraction to show how many fruitsicles are light colored, banana flavored.

     ———————

16. Mr. Smith's dog had 3 puppies. Two puppies are spotted and 1 puppy has solid colored fur. Write a fraction to show how many puppies have spots.

     ———————

17. There are 4 cupcakes on a plate. Write a fraction to show how many of the cupcakes have sprinkles on top.

     ———————

18. There are eight pairs of sandals. Write a fraction to show how many pairs have stars on the bows.

     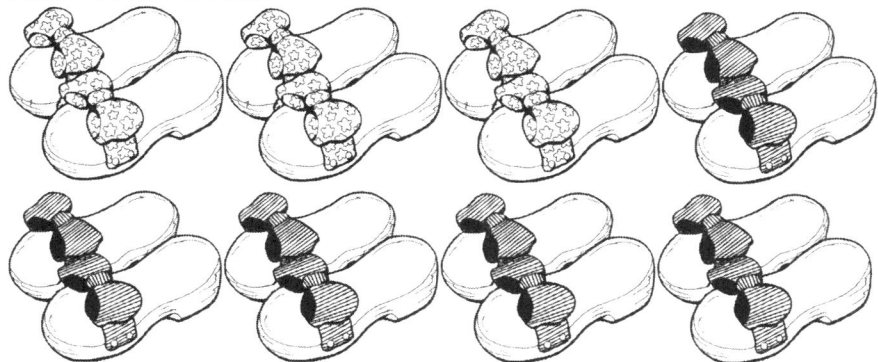

     ———————

19. There are 3 polar bears. All 3 polar bears are eating fish. Write a fraction to show how many polar bears are eating fish.

     ———————

20. There are 8 monkeys at the zoo exhibit. All 8 of the monkeys are playing. Write a fraction to show how many of the monkeys are playing.

     ———————

5.4 Fractions on a Number Line (DOK 2)

The space between each whole number on a number line is equal to one. Each space can be divided into parts to show fractions. The number of parts represents the denominator of the fraction. The dash that is plotted represents the numerator.

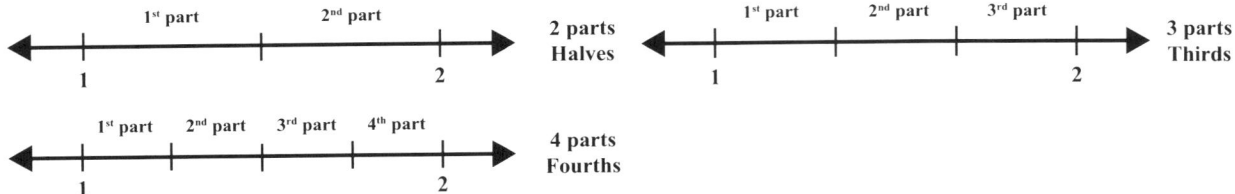

Read the examples below for more detail.

Example 1: Plot A at $\frac{1}{4}$ on the number line.

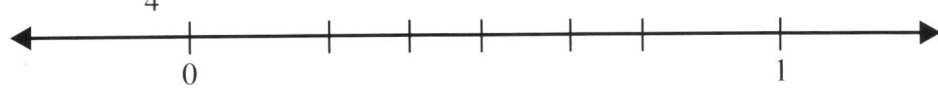

Step 1: Look at the denominator of the fraction. The denominator is 4. Divide the space between 0 and 1 into 4 parts. To do this, draw 3 equally spaced dashes.

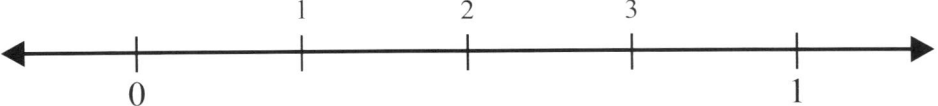

Look at the numerator of the fraction. The numerator is 1. Plot A on dash 1.

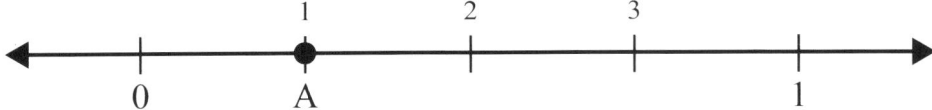

Example 2: Plot B at $\frac{2}{3}$ on the number line.

Step 1: Look at the denominator of the fraction. The denominator is 3. Divide the space between 0 and 1 into 3 parts. To do this, draw 2 equally spaced dashes.

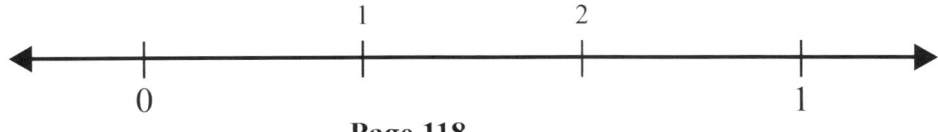

Step 2: Look at the numerator of the fraction. The numerator is 2. Plot B on dash 2.

Plot the points on the number lines provided. (DOK 2)

1. Plot K at $\frac{1}{2}$.

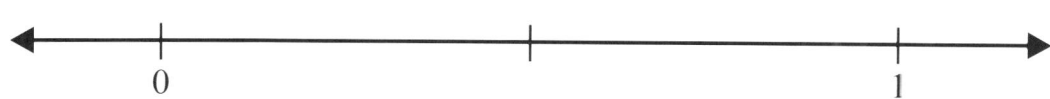

2. Plot L at $\frac{3}{4}$.

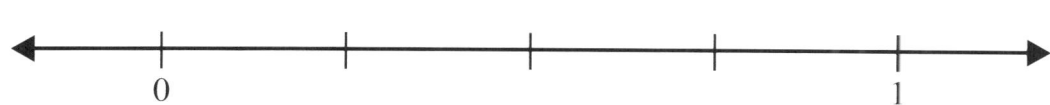

3. Plot M at $\frac{2}{3}$.

4. Plot N at $\frac{1}{4}$.

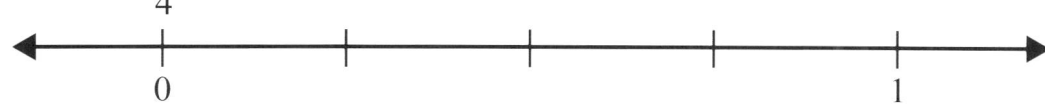

5. Plot P at $\frac{1}{3}$.

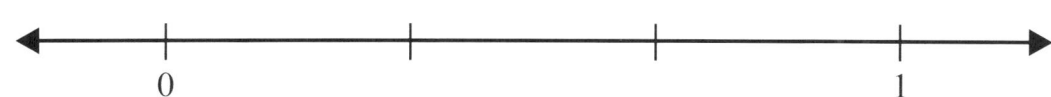

6. Plot Q at $\frac{5}{6}$.

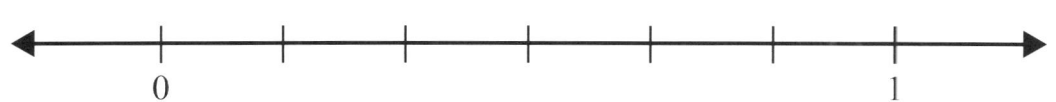

7. Plot R at $\dfrac{7}{8}$.

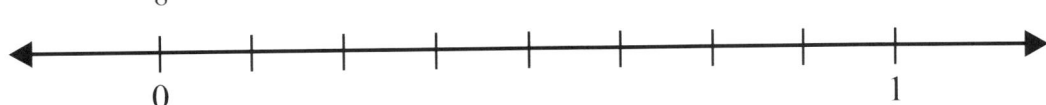

8. Plot S at $\dfrac{5}{9}$.

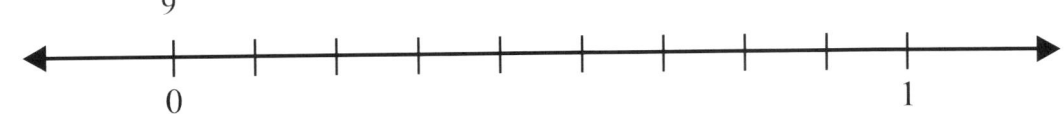

Choose the fraction that can replace the letter for each problem. (DOK 2)

9. Which fraction represents W on the number line?

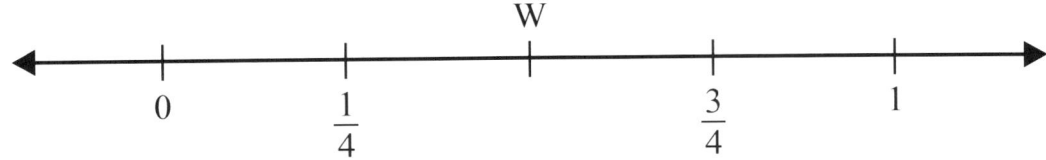

   A) $\dfrac{1}{3}$ B) $\dfrac{2}{4}$ C) $\dfrac{2}{3}$ D) $\dfrac{3}{8}$

10. Which fraction represents G on the number line?

   A) $\dfrac{1}{3}$ B) $\dfrac{1}{4}$ C) $\dfrac{3}{4}$ D) $\dfrac{1}{2}$

11. Which fraction represents D on the number line?

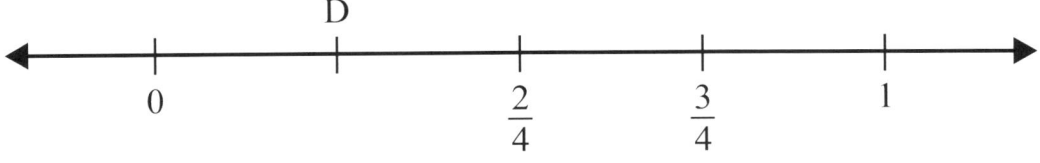

   A) $\dfrac{1}{3}$ B) $\dfrac{3}{4}$ C) $\dfrac{1}{4}$ D) $\dfrac{2}{3}$

12. Which fraction represents *T* on the number line?

A) $\frac{2}{3}$ B) $\frac{1}{2}$ C) $\frac{3}{4}$ D) $\frac{1}{4}$

13. Which fraction represents *C* on the number line?

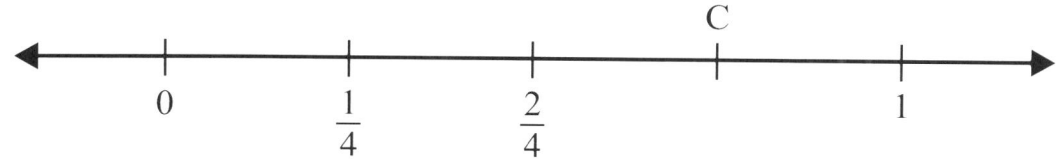

A) $\frac{1}{3}$ B) $\frac{5}{8}$ C) $\frac{2}{3}$ D) $\frac{3}{4}$

5.5 Comparing Fractions (DOK 2)

To compare fractions, plot each fraction on a number line. The fraction that is closer to 1 is bigger. The fraction that is closer to 0 is smaller.

Example 1: Compare $\frac{3}{6}$ and $\frac{3}{8}$.

    Step 1: Plot $\frac{3}{6}$ on the number line.

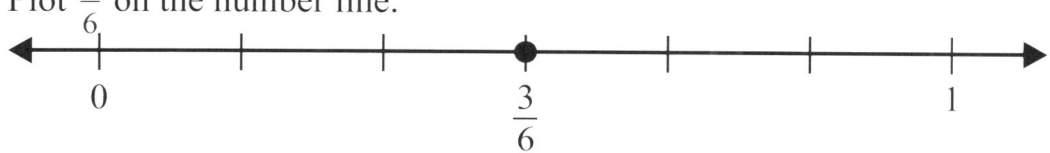

    Step 2: Plot $\frac{3}{8}$ on the number line.

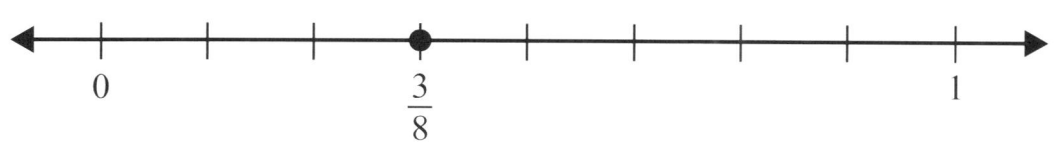

    Step 3: Compare the two number lines. Since $\frac{3}{6}$ is closer to 1, $\frac{3}{6} > \frac{3}{8}$.

Example 2: Compare $\frac{4}{7}$ and $\frac{5}{6}$.

Step 1: Plot $\frac{4}{7}$ on the number line.

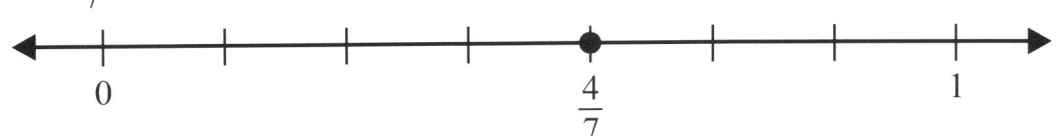

Step 2: Plot $\frac{5}{6}$ on the number line.

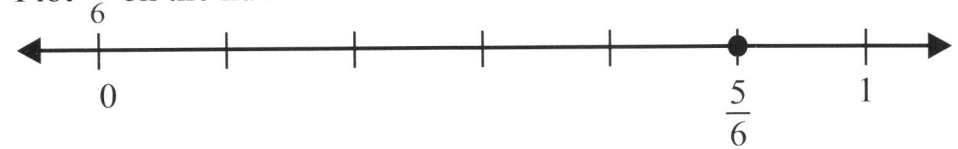

Step 3: Compare the two number lines. Since $\frac{5}{6}$ is closer to 1, $\frac{5}{6} > \frac{4}{7}$.

Fill in the boxes below with either < or >. (DOK 2)

1. $\frac{7}{8} \square \frac{2}{8}$

2. $\frac{1}{3} \square \frac{2}{3}$

3. $\frac{2}{4} \square \frac{1}{4}$

4. $\frac{3}{6} \square \frac{5}{6}$

5. $\frac{5}{8} \square \frac{2}{8}$

6. $\frac{3}{4} \square \frac{2}{4}$

7. $\frac{4}{4} \square \frac{1}{4}$

8. $\frac{2}{3} \square \frac{1}{3}$

9. $\frac{3}{4} \square \frac{1}{4}$

10. $\frac{3}{6} \square \frac{1}{6}$

11. $\frac{7}{8} \square \frac{8}{8}$

12. $\frac{1}{6} \square \frac{5}{6}$

Fill in the boxes below with either < or >. (DOK 2)

13. $\frac{2}{8} \square \frac{2}{6}$

14. $\frac{1}{4} \square \frac{1}{6}$

15. $\frac{3}{4} \square \frac{3}{8}$

16. $\frac{3}{4} \square \frac{3}{6}$

17. $\frac{5}{6} \square \frac{5}{8}$

18. $\frac{1}{4} \square \frac{1}{8}$

19. $\frac{2}{8} \square \frac{2}{4}$

20. $\frac{2}{3} \square \frac{2}{6}$

21. $\frac{1}{2} \square \frac{1}{3}$

22. $\frac{6}{8} \square \frac{6}{6}$

23. $\frac{4}{8} \square \frac{4}{6}$

24. $\frac{2}{6} \square \frac{2}{4}$

5.6 Modeling Fractions (DOK 2)

Fraction strips can be used to model fractions.

Example 1: Which of the fractions below is the smallest?
Which of the fractions below is the largest?

$$\frac{1}{4} = \boxed{\frac{1}{4}}$$

$$\frac{1}{3} = \boxed{\frac{1}{3}}$$

$$\frac{1}{2} = \boxed{\frac{1}{2}}$$

$$\frac{3}{4} = \boxed{\frac{1}{4}}\boxed{\frac{1}{4}}\boxed{\frac{1}{4}}$$

$$\frac{1}{8} = \boxed{\frac{1}{8}}$$

Step 1: Look at the length of each fraction strip.
Decide which is shortest and which is longest.

The shortest fraction strip is $\frac{1}{8}$.

The longest fraction strip is $\frac{3}{4}$.

Answer: The smallest fraction is $\frac{1}{8}$.

The largest fraction is $\frac{3}{4}$.

Answer each question about comparing fractions using models. (DOK 2)

1. Amy drew these fraction strips.

   $$\boxed{\frac{1}{2}} \quad \boxed{\frac{1}{8}} \quad \boxed{\frac{1}{3}}$$

   How should she put the 3 fraction strips in order from <u>smallest</u> to <u>largest</u>?

2. How many $\frac{1}{4}$ fraction strips are equal to a $\frac{3}{4}$ fraction strip?

3. Jonathan drew a shaded fraction of a rectangle to show $\frac{6}{8}$.

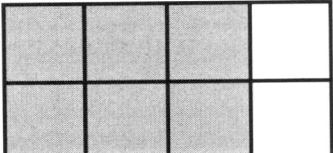

   Which of the rectangles below shows a shaded fraction larger than $\frac{6}{8}$?

   A) 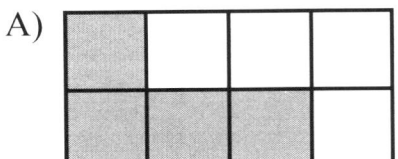   B)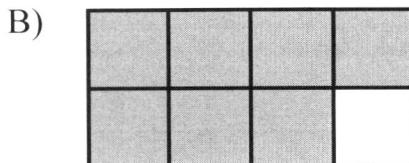

4. Below are three shaded fractions of rectangles.

   A) 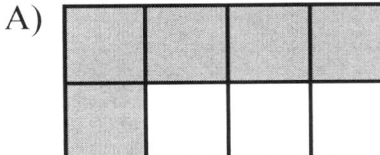   B) 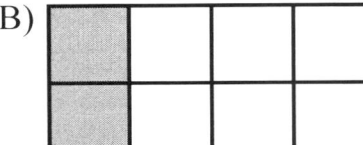   C)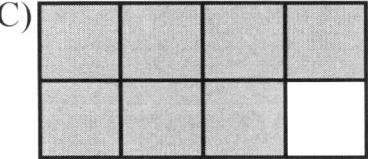

   How should these be placed so the 3 shaded fraction rectangles are in order from <u>smallest</u> to <u>largest</u>?

5. Which picture shows the <u>larger</u> fraction of speckled chicks? (Do not count the mother bird.)

   A)    B)

6. Which group of socks has the <u>smallest</u> fraction of this sock: ?

A) 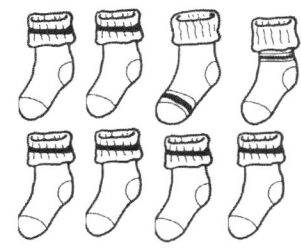   B)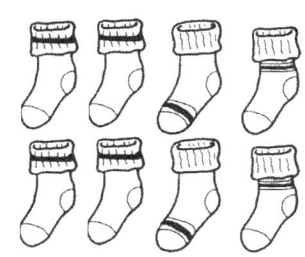

7. Which group shows a <u>larger</u> fraction of cupcakes with cherries on top?

A)    B)

8. Which group of dogs has the <u>larger</u> fraction of solid colored dogs?

A)    B)

5.7 Equivalent Fractions (DOK 3)

Equivalent fractions represent the same amount but can have different numbers in the numerator and denominator. The 3 models below show equivalent portions each with a different fraction.

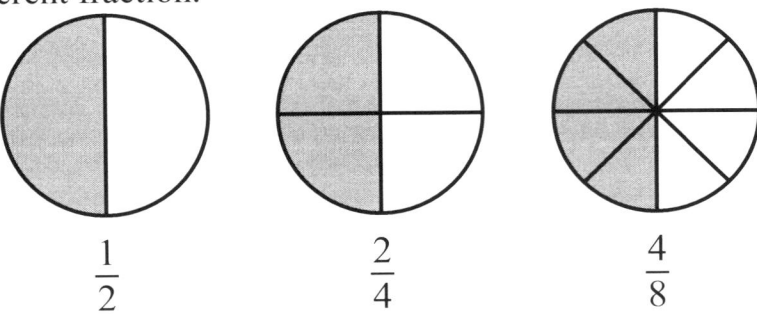

$$\frac{1}{2} \qquad\qquad \frac{2}{4} \qquad\qquad \frac{4}{8}$$

Example 1: A circle is divided into 3 parts. One part is shaded.

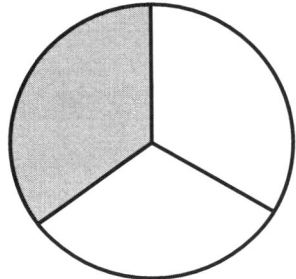

This picture represents $\frac{1}{3}$.

The circle is divided 3 more times.

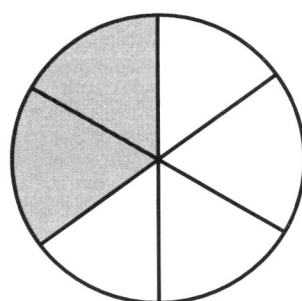

This picture represents $\frac{2}{6}$.

Since the same amount is shaded in both pictures, $\frac{1}{3}$ and $\frac{2}{6}$ are equivalent fractions.

Example 2: See the fraction strips and number line below.

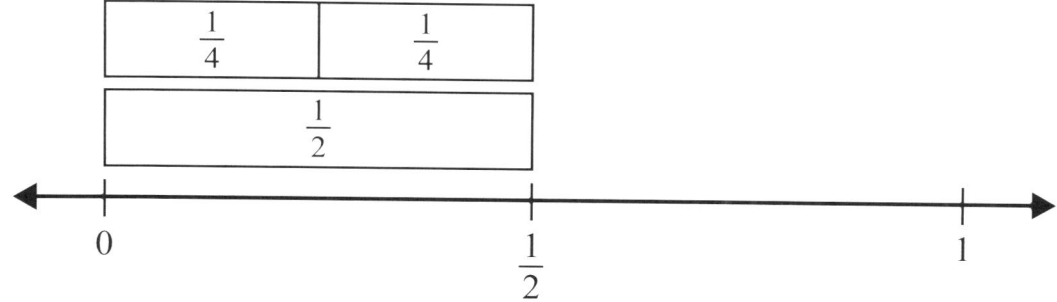

Two $\dfrac{1}{4}$ strips equal one $\dfrac{1}{2}$ strip.

Two $\dfrac{1}{4}$ strips $= \dfrac{1 \times 2}{4} = \dfrac{2}{4}$.

So $\dfrac{2}{4}$ and $\dfrac{1}{2}$ are equivalent fractions.

Read each problem and solve. (DOK 1)

1. Brendon divided a circle into 2 parts. He shaded 1 part.

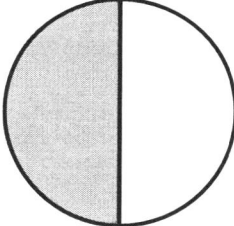

Then, he drew another line, as shown.

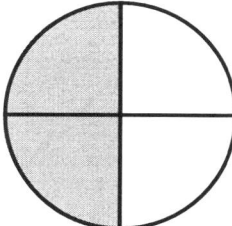

Fill in the box to show what fraction of the second model Brendon shaded.

$\dfrac{1}{2} = \dfrac{\square}{4}$

2. Ashley divided a hexagon into 3 parts. She shaded 1 part.

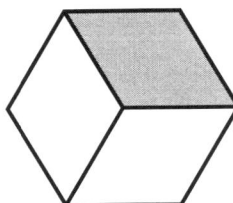

Then, she divided the same hexagon into 6 parts, as shown.

Fill in the box to show what fraction of the second model Ashley shaded.

$\dfrac{1}{3} = \dfrac{\boxed{}}{6}$

3. Roberto divided a square into 4 parts and shaded 1 part.

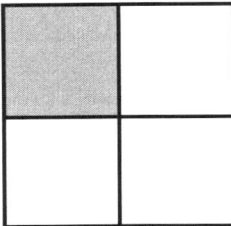

Then, he drew 2 more lines, as shown.

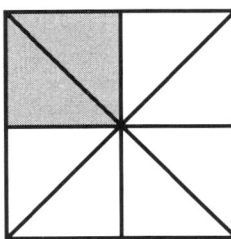

Fill in the box to show what fraction of the second model Roberto shaded.

$\dfrac{1}{4} = \dfrac{\boxed{}}{8}$

4. Janeen folded a piece of paper into 3 parts and shaded 2 parts.

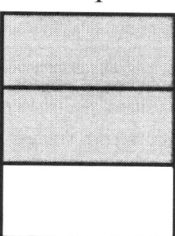

Then, she drew one more line, as shown.

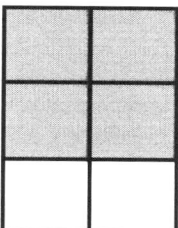

Fill in the box to show what fraction of the second model Janeen shaded.

$$\frac{2}{3} = \frac{\square}{6}$$

Match the model in column 1 to the equivalent model in column 2. The first one is done for you. (DOK 2)

Column 1   **Column 2**

5.      A)

6.      B)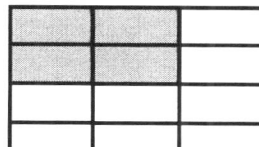

7.   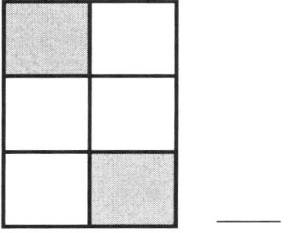   C)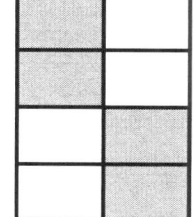

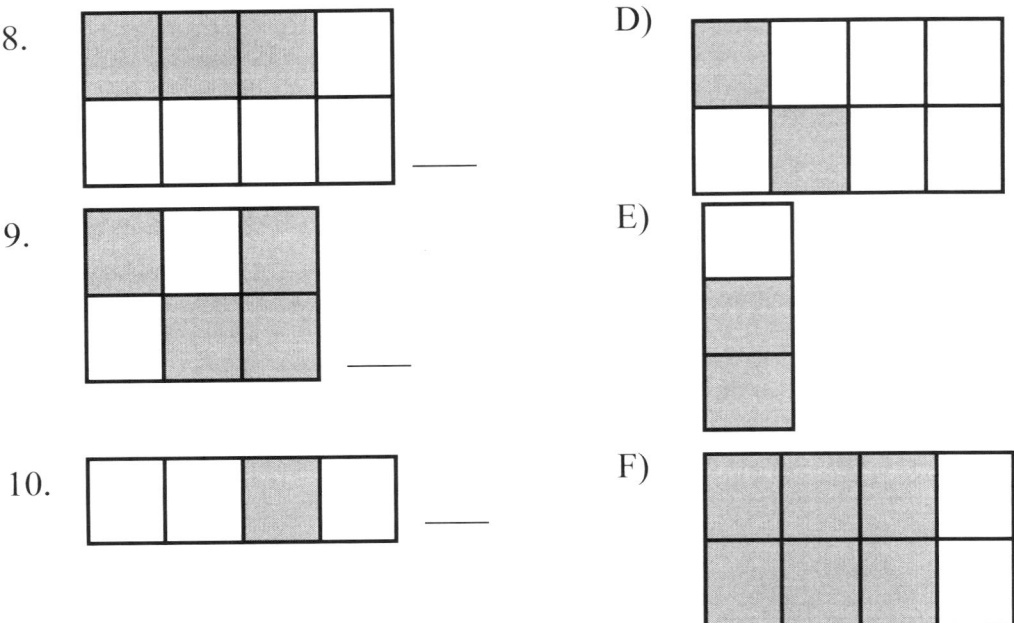

8. _____

9. _____

10. _____

D)

E)

F)

Use the number line and fraction strips below to answer the questions about equivalent fractions. (DOK 2)

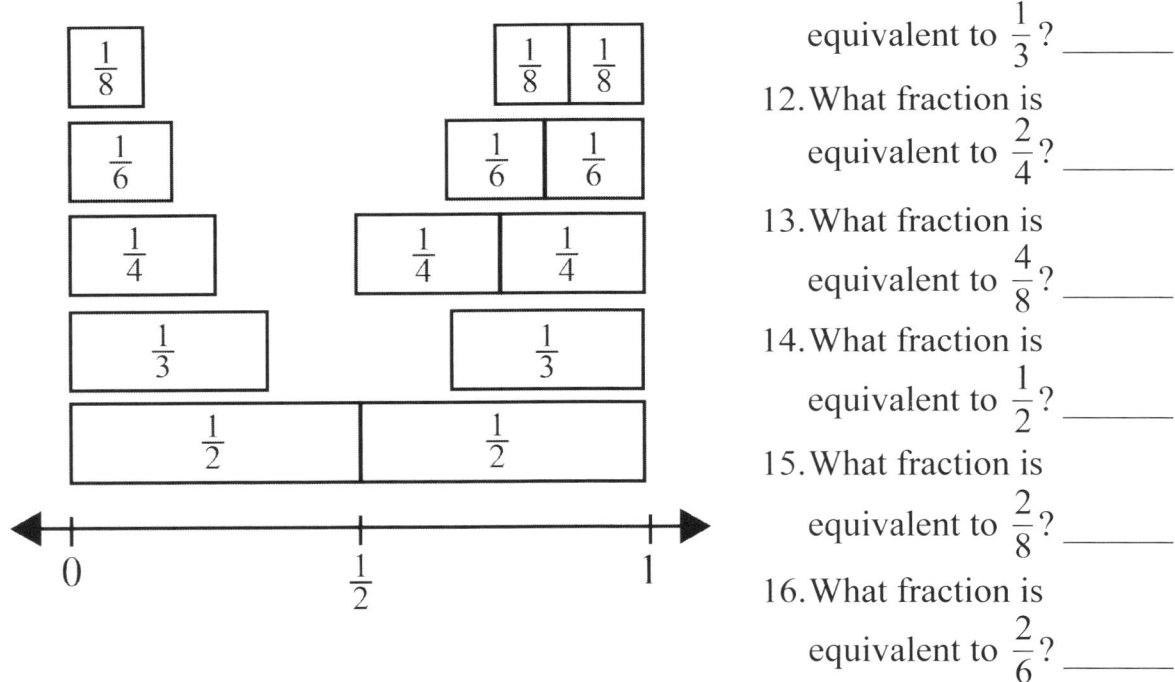

11. Which fraction is equivalent to $\frac{1}{3}$? _____

12. What fraction is equivalent to $\frac{2}{4}$? _____

13. What fraction is equivalent to $\frac{4}{8}$? _____

14. What fraction is equivalent to $\frac{1}{2}$? _____

15. What fraction is equivalent to $\frac{2}{8}$? _____

16. What fraction is equivalent to $\frac{2}{6}$? _____

5.8 Fractions Enrichment (DOK 2)

Solve the word problems. (DOK 2)

1. Lillie and her brother ate 3 of the 4 pieces of pie. What fraction of the pie did they eat?

    _____

2. Nelia's mother cut a pizza into 8 slices. Each of her 5 children ate 1 slice. What fraction of the pizza was <u>not</u> eaten?

    _____

3. Marcus must take 6 tests this year. He has taken 4 tests. What fraction of the tests has he taken?

    _____

4. Emily has 2 pairs of tennis shoes, 3 pairs of sandals, and 1 pair of dress shoes. What fraction of her shoes are tennis shoes?

    _____

5. Francis cut 8 paper snowflakes at school. She gave 5 to her father. What fraction of paper snowflakes did she give to her father?

    _____

6. Susan has 4 stuffed bears. She gives 2 of them to charity. What fraction did she give away?

    _____

7. Jacob has 6 video games. He puts 3 of them in a garage sale. What fraction of the video games does he put in the garage sale?

    _____

8. Tommy's mom makes 8 pancakes. His mom burned 3 pancakes. What fraction of the pancakes were <u>not</u> burned?

    _____

9. Matthew has 6 toy cars. Five of the toy cars are red. What fraction of the toy cars are red?

    _____

10. Eric has 2 action movies, 3 cartoon movies, and 1 funny movie. What fraction of his movies are funny movies?

_____

11. Madison made 8 snowballs. She threw 7 of the snowballs. What fraction of the snowballs did she throw?

_____

12. Brookside Elementary School has 8 swings on the playground. Six of the swings are being used. What fraction of the swings are being used?

_____

Look at the fraction models in column 1, and find the equivalent fraction models in column 2. (DOK 3)

13. _____

A)

14. _____

B)

15. _____

C)

16. _____

D)

17. 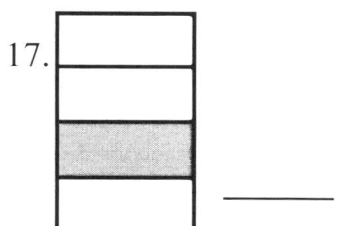 _____

E)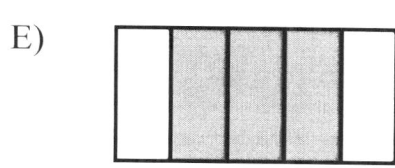

18. 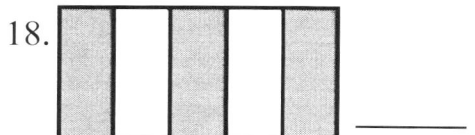 _____

F)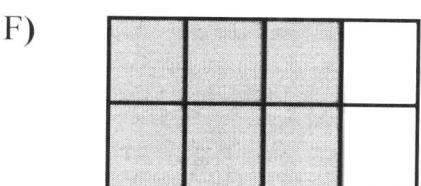

Chapter 5 Review

Find the fraction of the shaded part of each figure. (DOK 1)

1.  _____

3. 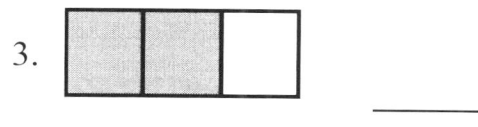 _____

2.  _____

4.  _____

Write the fractions using words. (DOK 1)

5. $\frac{2}{6}$ _____

6. $\frac{5}{8}$ _____

Compare the fractions. Use < , >, or = . (DOK 2)

7. $\frac{1}{3} \square \frac{2}{6}$

8. $\frac{1}{4} \square \frac{3}{4}$

9. $\frac{1}{6} \square \frac{1}{8}$

Answer the questions using the fraction strips and number line model. (DOK 2)

10. What fraction is equivalent to $\frac{1}{3}$? _____

11. What fraction is equivalent to $\frac{1}{2}$? _____

12. What fraction is equivalent to $\frac{2}{8}$? _____

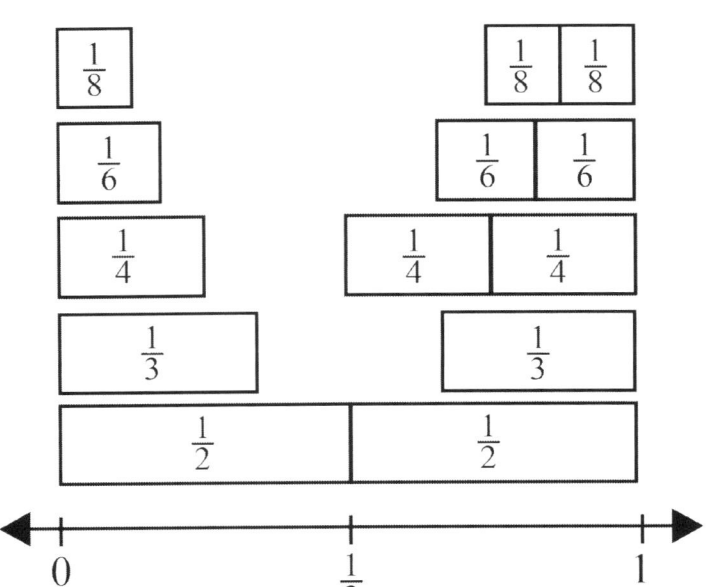

Answer each fraction question. (DOK 2)

13. Compare the shaded parts of the 2 fraction models below using > or <.

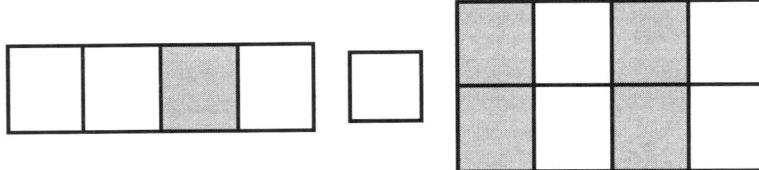

14. What fraction of the cupcakes have cherries on them?

15. Andrew has 4 dogs. All 4 dogs are outside playing in the snow. What fraction of the dogs are outside? _____

16. Mrs. Carson made 6 cakes for a bake sale. Two cakes have been sold. What fraction of the cakes have <u>not</u> been sold? _____

17. Which picture shows $\frac{2}{3}$ of the fish are facing to the right?

A)    B)

18. Which model shows $\frac{3}{4}$ bugs?

A) 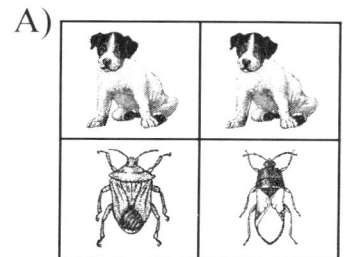   B) 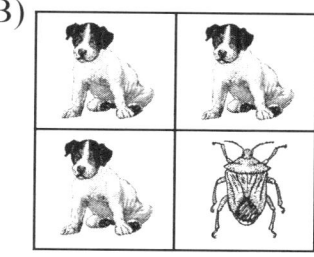   C)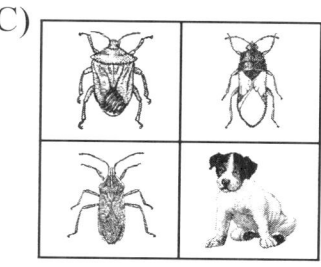

Follow the directions for each problem. (DOK 3)

19. Using the figures below, fill in $\frac{6}{10}$ of the stars, circle $\frac{3}{10}$ of the stars, and do nothing to $\frac{1}{10}$ of the stars.

20. Shade $\frac{3}{4}$ of the boxes below.   21. Shade $\frac{2}{5}$ of the boxes below.

Choose the correct answer for each problem below. (DOK 3)

22. Amy is reading a book. She has read the same fraction of the book as the shaded model below.

What fraction of the book has she read?

A) $\frac{2}{3}$ B) $\frac{3}{4}$ C) $\frac{1}{4}$ D) $\frac{2}{4}$

23. There are 12 eggs in a dozen. The shaded model below shows how many eggs are left in the dozen.

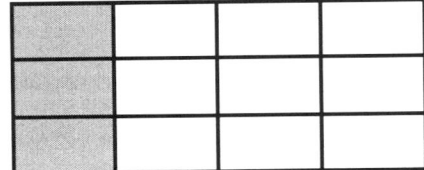

What fraction of the eggs are left in the dozen?

A) $\frac{3}{12}$ B) $\frac{6}{12}$ C) $\frac{8}{12}$ D) $\frac{4}{12}$

24. The shaded model below shows the portion of strawberries that have been eaten.

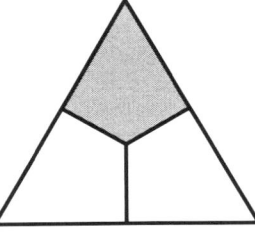

What fraction of the strawberries have been eaten?

A) $\frac{1}{4}$ B) $\frac{1}{2}$ C) $\frac{2}{3}$ D) $\frac{1}{3}$

25. The fraction model below shows the portion of sandals with striped bows.

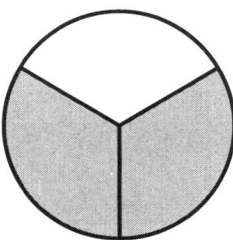

Which fraction of the sandals have striped bows?

A) $\frac{1}{3}$ B) $\frac{1}{4}$ C) $\frac{1}{2}$ D) $\frac{2}{3}$

For additional practice, please see Chapter 5 Test located in the Teacher Guide.

Chapter 6
Measurement

This chapter covers the following Grade 3 standards:

	Content Standard
Measurement and Data	3.MD.1, 3.MD.2, 3.MD.4

6.1 Telling Time (DOK 1)

We can tell time using an analog clock or a digital clock. An **analog clock** has an hour hand, minute hand, and second hand located on the face of a clock as shown below.

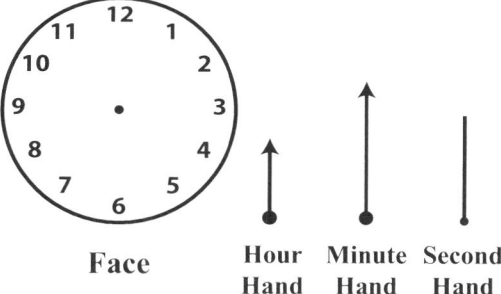

There are 60 seconds in a minute and 60 minutes in an hour.

A **digital clock** displays the same information as an analog clock. The digital clock lists the time from left to right.

Digital clocks also list a.m. and p.m. There are 24 hours in a day. AM describes the first 12 hours of the day (sunrise/morning). PM describes the second 12 hours of the day (sunset/evening).

Example 1: The clocks below display the same time.

Each clock shows 3:25:15. The digital clock specifies morning (a.m.) or afternoon (p.m.).

Example 2: Rebecca wakes up for school at the same time every morning and looks at the clock. What time does Rebecca wake up? Use the analog clock to fill in the digital clock.

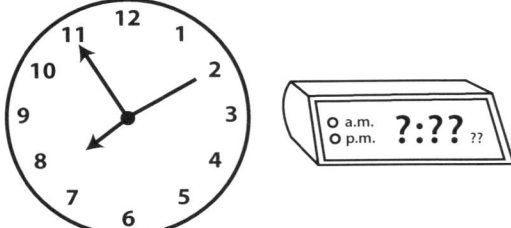

Step 1: Look at the short hand (hour hand). What two numbers is it between?

It is pointing between 7 and 8. That means it is hour 7.

Step 2: Look at the long hand (minute hand). What number is it closest to?

It is pointing to the 11. Multiply 11 by 5.

$11 \times 5 = 55$

So it is 55 minutes after 7, or 7:55.

Page 138

Step 3: Fill in the digital clock.

The 7 is placed to the left of the dots. The 55 is placed to the right of the dots. Since it is the morning, fill in the a.m. dot.

Write the time below. (DOK 1, 2)

1.

_____

5.

_____

2.

_____

6.

_____

3. Draw the time on the analog clock provided.

4. Draw the time on the analog clock provided.

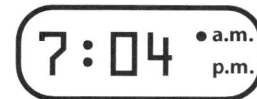

7. Draw the time on the analog clock provided.

8. Draw the time on the analog clock provided.

For questions 9–12, decide if the scenario occurs in the first half of the day or the second half of the day. Circle a.m. or p.m.

9. Sammy and Olivia catch lightning bugs after dinner.   a.m.   p.m.

10. Mary Jane eats pancakes for breakfast.   a.m.   p.m.

11. Carolina watches the sunset over the lake.   a.m.   p.m.

12. Georgio brushes his teeth before school.   a.m.   p.m.

6.2 Elapsed Time (DOK 2)

Elapsed time is the amount of time that passes from the beginning of an event to the end of that event.

Example 1:   Christopher has to feed his puppy at 1:30 p.m.
How long until Christopher feeds his puppy?

Step 1:   Look at the clock. It currently reads 12:00 p.m.

Step 2:   Count the number of hours that will pass.
Start time: 12:00 p.m.
End time: 1:30 p.m.

One hour passes between the start time and end time. After 1 hour it is 1:00 p.m.

Step 3:   Count the number of minutes that will pass.

Start time: 1:00 p.m.
End time: 1:30 p.m.

Page 140

Thirty minutes pass between the next start time and end time.

Step 4: Add the hours and minutes together.

Answer: Christopher will feed his puppy in 1 hour and 30 minutes.

Example 2: Recess begins at 1:45 p.m. and ends at 2:30 p.m. How much recess time do the students have?

Count using 15-minute intervals:

1:45	–	2:00	=	15 minutes
2:00	–	2:15	=	15 minutes
2:15	–	2:30	=	15 minutes
				45 minutes

Answer: The students have 45 minutes for recess.

Answer the questions below. Show your work. (DOK 2)

1. Henry's baseball practice starts at 2:00 p.m. It lasts 1 hour and 45 minutes. What time does he get out of baseball practice?

   _____

2. Shasta is going to her best friend's house for dinner. Her mother says she needs to be back in 2 hours and 30 minutes. If she leaves at 5:00 p.m., what time must she be back?

   _____

3. Jayden and Jacques went to the park. They were gone for 1 hour and 15 minutes. They got home at 12:30 p.m. What time did they leave for the park?

   _____

4. Valerie and her family are going to her grandparent's house. It is a 3-hour drive. If they leave at 9:00 a.m. without stopping, what time will they be at Valerie's grandparent's house?

   _____

Clock A

Clock B

Clock C

Clock D

Clock E

Clock F

5. Reading class started at 9:05 a.m. It ended 32 minutes later. Which clock shows the time it ended?

_____

6. Lunch is served at 11:45 a.m. If Kevin has to wait 11 minutes until lunch is served, what time is it right now?

_____

7. Jamille only has 3 minutes left before she goes to bed at 9:00 p.m. Which clock shows what time it is now?

_____

8. Christy's mom gets off work at 5:00 p.m. In 6 minutes, Christy's mom will leave work. Which clock shows what time it is now?

_____

9. Andy gets out of school at 3:00 p.m. Andy left school 3 minutes ago. Which clock shows what time is it now?

_____

10. Mr. Watson's gym class started at 2:00 p.m. Thirty-four minutes have passed. Which clock shows the current time?

_____

6.3 Time Intervals (DOK 2)

Time intervals can be shown on a number line. Going from left to right on the number line is the same as going from earlier time to later time. The time interval number line below shows 11:00 a.m. to 1:00 p.m.

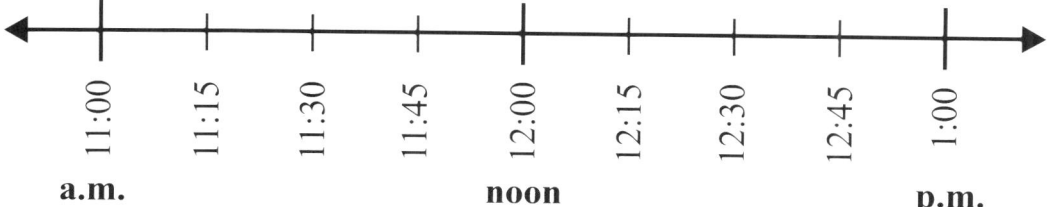

Each of the quarter hours is marked off on the number line: 11:15, 11:30, 11:45, and so on. Each quarter of an hour is equal to 15 minutes.

Example 1: It is 11:00 a.m., and John goes to lunch at school at 11:45 a.m. How long does John have to wait for lunch?

Step 1: Find 11:00 a.m. on the number line above.

Step 2: Find 11:45 on the number line above, and count the quarter-hour spaces between 11:00 a.m. and 11:45 a.m. There are 3 spaces. Three quarters × 15 minutes = 45 minutes

Answer: John has to wait 45 minutes until lunch.

Example 2: It is 11:30 a.m., and Mia will go outside for recess at 12:45 p.m. How long does Mia have to wait for recess?

Step 1: Find 11:30 a.m. on the number line above.

Step 2: Find 12:45 on the number line above, and count the quarter-hour spaces between 11:30 a.m. and 12:45 p.m. There are 5 quarter-hour spaces.
Five quarters × 15 minutes = 1 hour and 15 minutes
(There are 4 quarters in each hour.)

Answer: Mia has to wait 1 hour and 15 minutes until recess.

Answer the two questions that follow each time interval number line. (DOK 2)

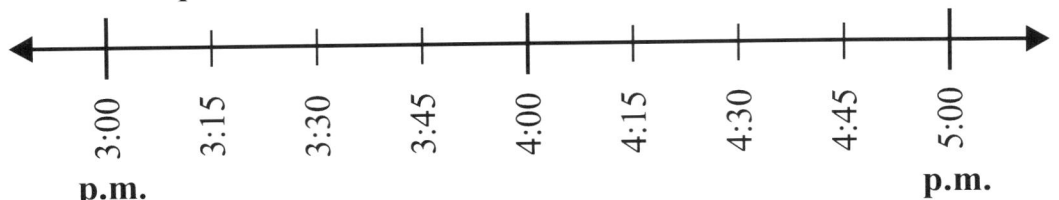

1. Julia left school at 3:00 p.m. She got home at 3:30 p.m. How long did it take Julia to get home?

_____

2. Maria started her homework at 3:45 p.m. and finished at 5:00 p.m. How long did it take to do her homework?

_____

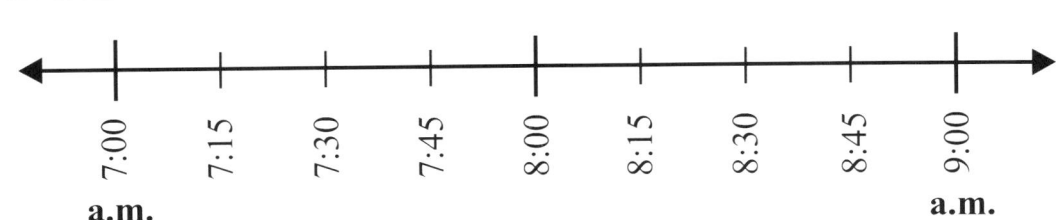

3. Ashley went walking with her mother at 7:15 a.m. They returned home at 8:30 a.m. How long did they walk?

_____

4. Ethan left his house at 7:45 a.m. He arrived at his friend's house at 8:00 a.m. How long did it take Ethan to get to his friend's house?

_____

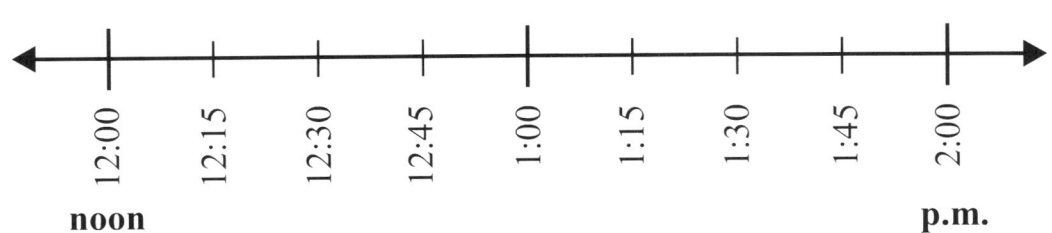

5. Lunch is at noon. Recess is right after lunch. Recess is over at 1:30 p.m. How long are lunch and recess all together?

_____

6. Janine started her chores on Saturday at 12:45 p.m. She finished at 2:00 p.m. How long did it take Janine to do her chores?

_____

6.4 Mass (DOK 2)

Mass is how much something weighs. It is measured using grams and kilograms.

Mass		Approximation
gram (g)	smallest unit	about the weight of a large paperclip
kilogram (kg)	1,000 g = 1 kg	a little over 2 pounds

Types of objects measured in grams: baking flour, thumb tacks, cereal, dollar bills

Types of objects measured in kilograms: animals, humans, furniture, books

When you estimate **mass**, your answer may not be exact. An estimate is a measurement close to the true mass of an object.

Example 1: Select the best estimate for the mass of a car.

1,000 g or 1,000 kg

Since a car is fairly heavy, we would choose 1,000 kg to estimate its mass.

Example 2: Select the best estimate for the mass of a cup of sugar.

100 g or 100 kg

Since a cup of sugar is a fairly small amount, we would choose 100 g to estimate its mass.

Which unit of measure would you use to estimate the mass of the following objects? Write grams or kilograms. (DOK 2)

1. Coffee _____

2. Iron filings _____

3. Meat _____

4. Truck tires _____

5. A handful of pebbles _____

6. Paper _____

7. Isabella has a package of cookies that weighs 425 grams. She serves her friends 47 grams of the cookies. How many grams of cookies are left in the package?

   _____

8. A box of dolls weighs 10 kilograms. There are 5 dolls in the box. What is the mass of each doll?

   _____

9. Nathan's dog had 7 puppies. At 3 months, the puppies each weigh about 14 kilograms. How many kilograms do the puppies weigh all together?

   _____

10. Rosa weighs 85 kilograms. Her mother weighs 323 kilograms. What is the difference in weight between Rosa and her mother?

    _____

11. A box of Fluffy-O's cereal weighs 400 grams. Darius and his brother together ate 70 grams of the cereal for breakfast. How many grams of cereal are left in the box?

    _____

6.5 Volume (DOK 2)

Volume is how much a container can hold. Volume is measured in liters and milliliters.

Volume	Approximation
milliliter (mL)	one-fifth of a teaspoon
liter (L)	about 4 glasses of water

When you **estimate volume**, your answer may not be exact. An estimate is a measurement close to the true volume.

Example 1: Select the best estimate for the volume of a large glass of milk.

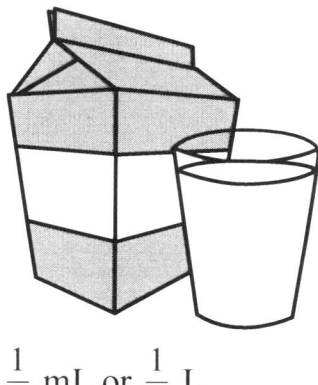

$\frac{1}{4}$ mL or $\frac{1}{4}$ L

Milliliters would be too small to measure a glass of milk. The best estimate for a large glass of milk would be $\frac{1}{4}$ L.

Example 2: Select the best estimate for the volume of medicine in a small syringe.

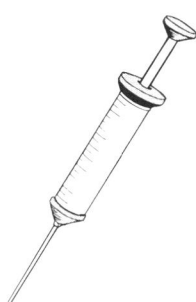

1 mL or 1 L

A syringe does not hold very much liquid. Milliliters would be the best choice for estimating the amount of medicine in a syringe.

Which unit of measure would you use to estimate the volume of the following objects? Write milliliters or liters. (DOK 2)

1. The amount of gasoline in a car _____

2. One rain drop _____

3. Large bottle of soda _____

4. The amount of nail polish in 1 bottle _____

5. Water in a swimming pool _____

6. The amount of ink in a pen _____

Solve. (DOK 2)

7. Samantha put 6 liters of water on one picnic table, 3 liters of water on another picnic table, and 5 liters of water on a third picnic table. How many liters of water did Samantha put on the 3 picnic tables?

    _____

8. Janice has a cold. She has to take 6 milliliters of cough syrup in the morning. She has to take 12 milliliters of cough syrup in the evening. How many milliliters does Janice have to take each day?

    _____

9. Mrs. Evans has a 4-liter container of sour cream at her bakery. She bakes a cake using 1 liter of sour cream. How many liters of sour cream are left in the container?

    _____

10. Mr. Tifton made 16 liters of barbecue sauce for his company party. He used 5 liters on barbecued chicken and barbecued ribs. How many liters of sauce are left over?

    _____

11. Ms. Albert has 16 liters of punch to share with her class on the last day of school. To make it easier to pour the punch, she divides the punch into 8 pitchers. How many pints of punch does each pitcher hold?

    _____

6.6 Length (DOK 2)

Length is how long something is. We can measure length using **standard** measurements. Standard measurement includes **inches**, **feet**, and **yards**. We can also use the **metric system**. The metric system uses **centimeters** and **meters**. When measuring, stay in the same unit of measure (standard or metric). Do not cross between the two systems of measure.

Standard Measurements

	Abbreviation	Equivalents	Used to Measure
Inch	in		small things: pets, books, babies
Foot	ft	12 inches = 1 foot	medium things: people, rooms
Yard	yd	3 feet = 1 yard	big things: football fields, parks

Metric Measurements

	Abbreviation	Equivalents	Used to Measure
Centimeter	cm		small things: pencils, bugs, paper
Meter	m	100 cm = 1 m	large things: tracks & fields

When measuring to the nearest $\frac{1}{2}$ inch or $\frac{1}{4}$ inch, follow the same rules as when measuring to the nearest inch. Instead of rounding up or down to the nearest whole number, round to the nearest $\frac{1}{2}$ inch or $\frac{1}{4}$ inch.

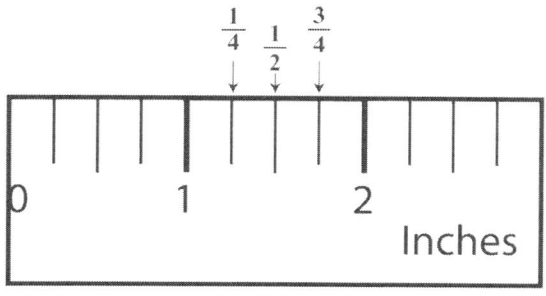

Example 1: What is the length of this paper clip to the nearest $\frac{1}{4}$ inch?

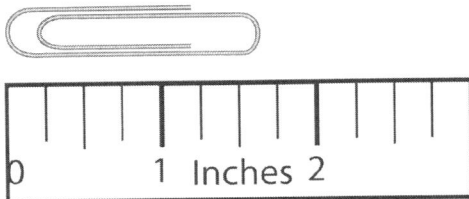

Answer: The paper clip is about $1\frac{3}{4}$ inches long.

Example 2: What is the height of this robot to the nearest $\frac{1}{2}$ inch?

Answer: The robot is about $1\frac{1}{2}$ inches tall (with antennae).

Using <u>standard</u> measurements, which unit would you use to measure the following items? Write inch, foot, or yard. (DOK 2)

1. soccer field   _____

2. length of a table   _____

3. crayon   _____

4. height of a tall man   _____

5. height of a shoe box   _____

6. length of a puppy   _____

Using <u>metric</u> measurements, which unit would you use to measure the following items? Write centimeter or meter. (DOK 2)

7. length of a grasshopper   ____

8. length of a track for running   ____

9. width of a piece of paper   ____

10. length of your thumb   ____

11. length of an airplane   ____

12. width of a game board   ____

Measure these toy items to the nearest $\frac{1}{2}$ inch. (DOK 2)

13.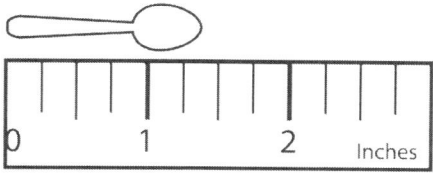

14.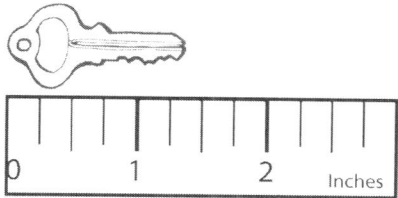

15.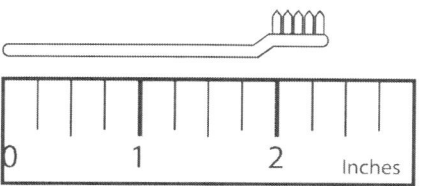

16.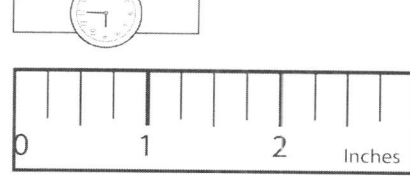

Measure to the nearest $\frac{1}{4}$ inch.

17.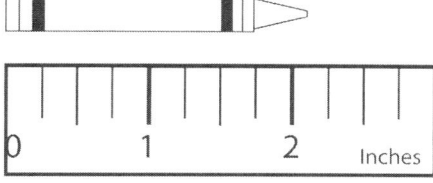

19.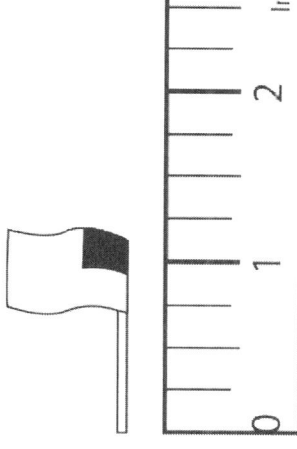

18.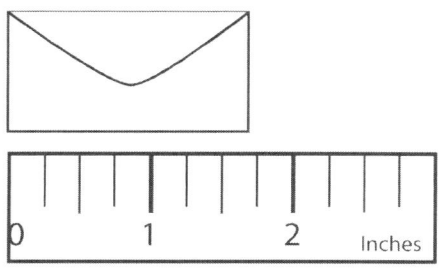

20.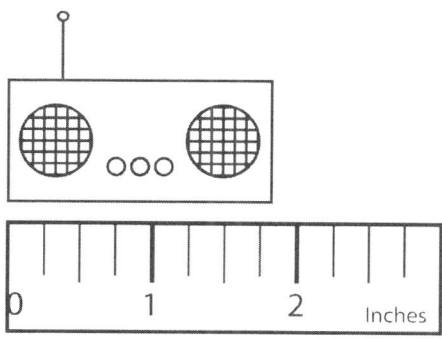

Chapter 6 Review

Answer the questions about time. (DOK 1, 2)

1. What time is shown?

_____

2. Paula went to the library at 5:45 p.m. If she left the library at 7:15 p.m., how long did she stay?

_____

3. What time is shown?

_____

4. Kathy's muffins will be ready to be taken out of the oven 20 minutes after the time shown. What time will they be ready?

_____

Use this time interval number line to answer the next two questions. (DOK 2)

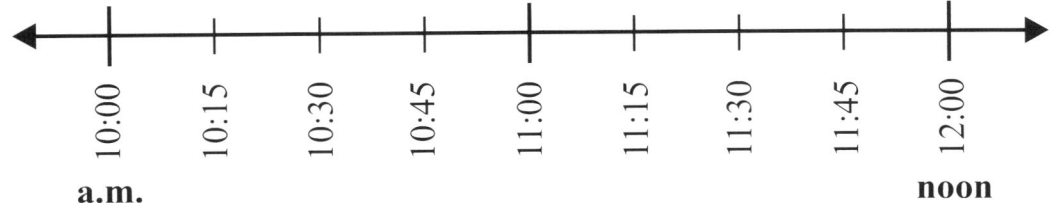

5. Alicia went to the park at 10:15 a.m. She returned home at 11:30 a.m. How long was Alicia at the park?

_____

6. Rick has a dentist appointment at 10:45 a.m. He leaves the dentist office at 11:30 a.m. How long was Rick at the dentist office?

_____

Answer the questions below.

7. It is 6:30 p.m. Jim's mother tells him he has to go to bed in 1 hour and 30 minutes. What time does Jim have to go to bed?

_____

8. Tracy left a birthday party at 3:00 p.m. She arrived at the party two hours earlier. What time did Tracy arrive at the birthday party?

_____

Answer the questions. (DOK 2, 3)

9. What would you use to measure the mass of a sparrow: a gram or a kilogram?

   _____

10. How would you measure the height of a school: by the yard or by the inch?

    _____

11. Is the volume of a bottle of sunscreen closer to 345 mL or 345 liters?

    _____

12. What would you use to measure the mass of a boiled egg: grams or kilograms?

    _____

13. There are 325 grams in a loaf of Yummy Cinnamon Bread. Lisa and Lily eat 37 grams of the bread. How many grams of bread are left?

    _____

14. Is the height of a reading book closer to 28 centimeters or 28 inches?

    _____

15. Aaron has 40 feet of rope. He divides the rope into 5 equal parts to make jump ropes. How long is each jump rope?

    _____

16. Is the volume of a glass of orange juice closer to 30 mL or 30 liters?

    _____

17. What would you use to measure the mass of a baby dinosaur: grams or kilograms?

    _____

18. Beau has 12 grams of flour, 15 grams of sugar, and 5 grams of salt. How many total grams of baking materials does Beau have?

    _____

19. A bottle of shampoo has 210 mL. Greta has used 15 mL of shampoo. How many milliliters of shampoo does Greta have left?

    _____

20. Measure the tree to the nearest inch.

_____

21. Measure the toy truck to the nearest half inch.

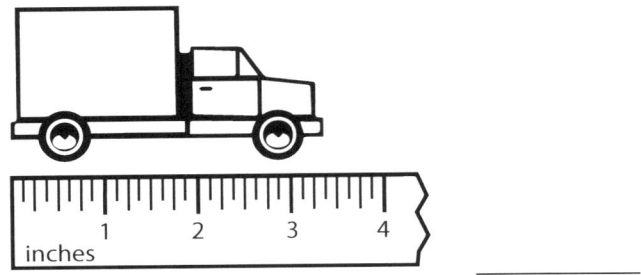

_____

22. Measure the toy football to the nearest inch.

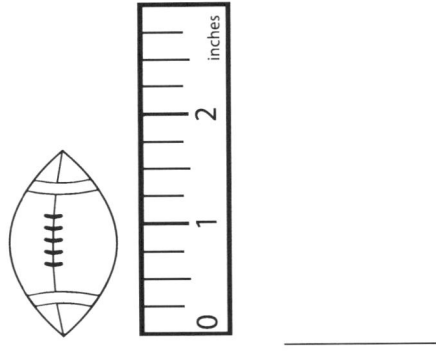

_____

23. Measure the key to the nearest cm.

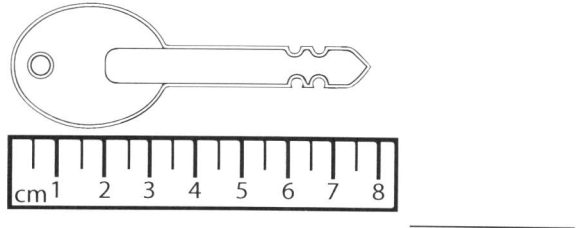

_____

24. Measure the bottle of glue to the nearest cm.

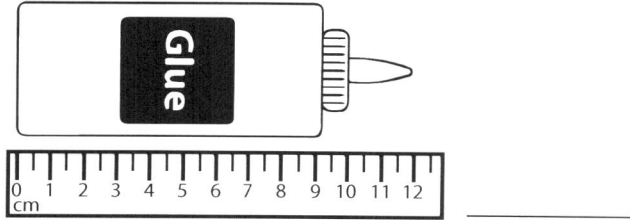

_____

25. Measure the toy duck to the nearest cm.

_____

26. Measure the plastic dog to the nearest inch.

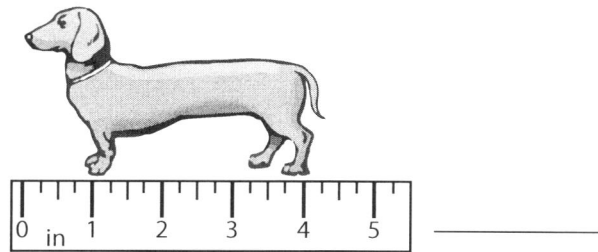

_____

For additional practice, please see Chapter 6 Test located in the Teacher Guide.

Chapter 7
Graphs

This chapter covers the following Grade 3 standards:

	Content Standard
Measurement and Data	3.MD.3, 3.MD.4

7.1 Bar Graphs (DOK 2)

A **bar graph** shows how items in a data set are organized and classified. The **category** is the information in the data set that is <u>not</u> described by a number. The **frequency** is the information in the data set that <u>is</u> described by a number. Each bar represents a category. The height of the bar shows the frequency of the category.

Example 1:   Nelson, Kela, Somu, and Jen collect basketball cards. Nelson has 4 cards, Kela has 6 cards, Somu has 7 cards, and Jen has 8 cards. Draw a bar graph that describes the data.

Step 1:   Identify the category. The category is the information in the data that is <u>not</u> described by a number. In this example, the **names** of the kids describe the category. Each name has 1 bar. This information is shown on the **x-axis**.

Step 2:   Identify the frequencies. The **frequency** is the information in the data set that <u>is</u> described by a number. The **number of cards** that each child has describes frequency. This information is shown on the **y-axis**.

Answer:

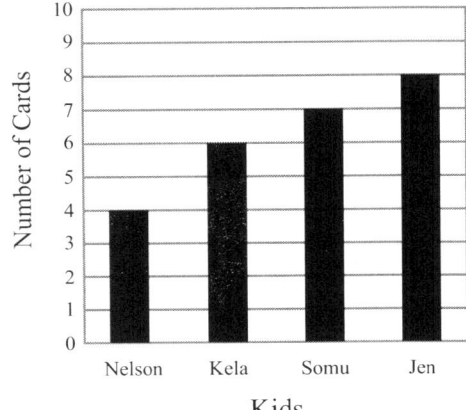

Page 156

Example 2: Sheeba asks 10 third-graders where their favorite place to swim is. She made a bar graph to describe the data. How many third-graders like swimming at the pool the most?

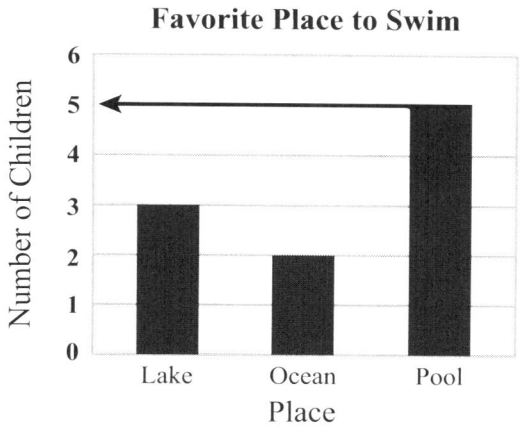

Favorite Place to Swim

Step 1: Look at the category labeled "Pool" on the *x*-axis.

Step 2: Find the value along the *y*-axis that lines up with the top of the "Pool" category. The height of the bar is 5.

Answer: Five third-graders like swimming at the pool the most.

Use the bar graph to answer the questions. (DOK 2)

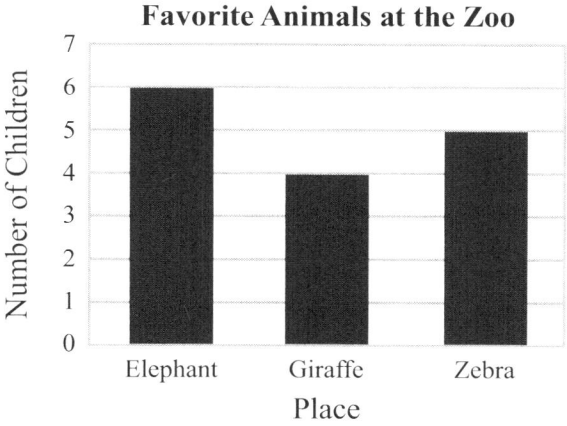

Favorite Animals at the Zoo

1. Which is the favorite animal to see at the zoo? _____

2. How many children like the giraffe the most? _____

3. How many more children like elephants than zebras? _____

4. How many children in total like elephants and zebras the most? _____

Use the bar graph to answer the questions. (DOK 2)

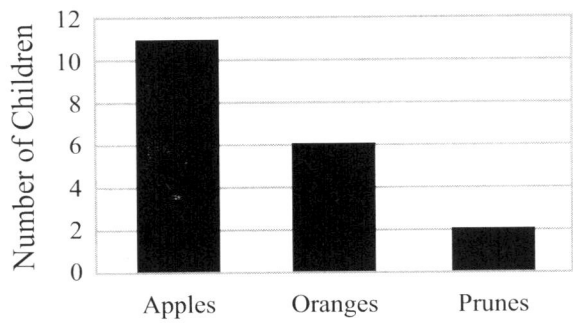

5. How many children said apples were their favorite fruit? _____

6. How many children said prunes were their favorite fruit? _____

7. How many more children said apples were their favorite fruit than children who said oranges were their favorite fruit? _____

8. How many children are represented in the bar graph in all? _____

9. In a 3rd grade class, 2 students like tag, 7 like soccer, 5 like baseball, and 4 like board games. Make a bar graph using the data in the table. (DOK 3)

Favorite Game	Number of Students
Tag	2
Soccer	7
Baseball	5
Board Games	4

7.2 Line Graphs (DOK 2)

Line graphs show how data changes over time. The amount of time is placed on the x-axis, and the frequency for each period of time is placed on the y-axis.

Example 1: Dora ate 4 cookies on Sunday, 1 cookie on Monday, 6 cookies on Tuesday, 9 cookies on Wednesday, 2 cookies on Thursday, 3 cookies on Friday, and 4 cookies on Saturday. Draw a line graph that describes the data.

Step 1: Place the days of the week on the x-axis. Plot a point to show how many cookies were eaten each day.

Step 2: Draw a line that connects all of the points.

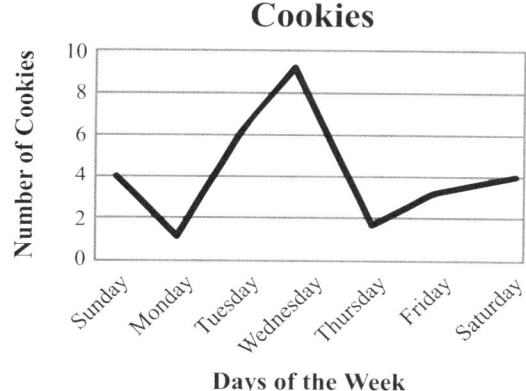

Example 2: The line graph below shows how many puppies were brought to the Friendly Pet Clinic from January to June. How many puppies were brought to the clinic in January? How many puppies were brought to the clinic in April? In what month were 50 puppies brought to the clinic?

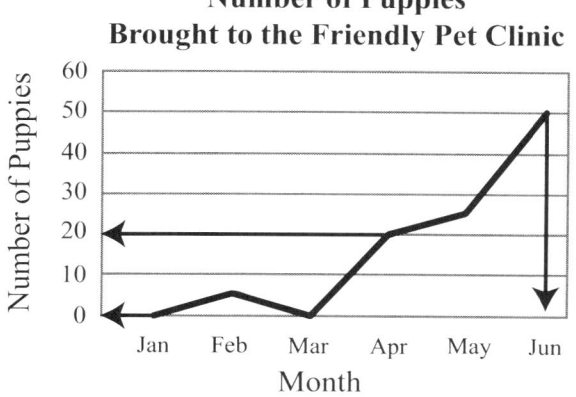

Step 1: How many puppies were brought to the clinic in January? Find January on the *x*-axis. Look across to the number on the *y*-axis. Since the line graph is marked at 0 on the *y*-axis, no puppies were brought to the clinic in January.

Step 2: How many puppies were brought to the clinic in April? Find April on the *x*-axis, and then follow the line across to the *y*-axis. Since the line graph is marked at 20 on the *y*-axis, 20 puppies were brought to the clinic in April.

Step 3: In what month were 50 puppies brought to the clinic? Find 50 on the *y*-axis. Follow this marker over to the line graph to find the month. Since 50 lines up with the line at the month of June, 50 puppies were brought to the clinic in June.

Use the line graph below to answer the questions. (DOK 2)

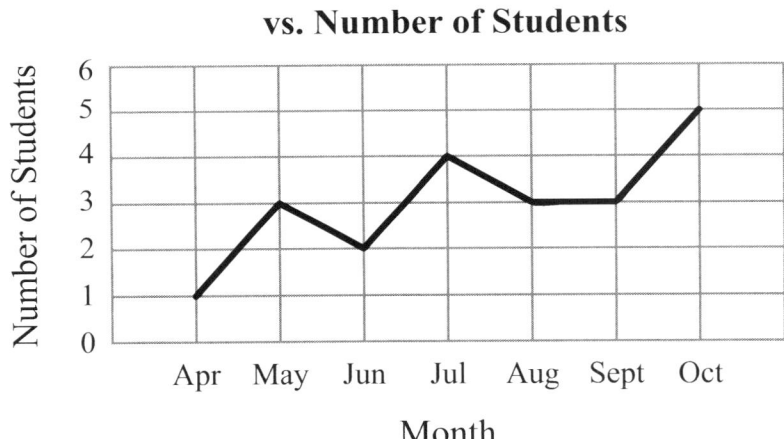

1. Which month has 2 birthdays in Mrs. Neville's 3rd grade class? _____

2. How many students have birthdays in October? _____

3. Which three months have 3 birthdays in them? _____

4. How many students have birthdays in July? _____

Use the line graph below to answer the questions. (DOK 2)

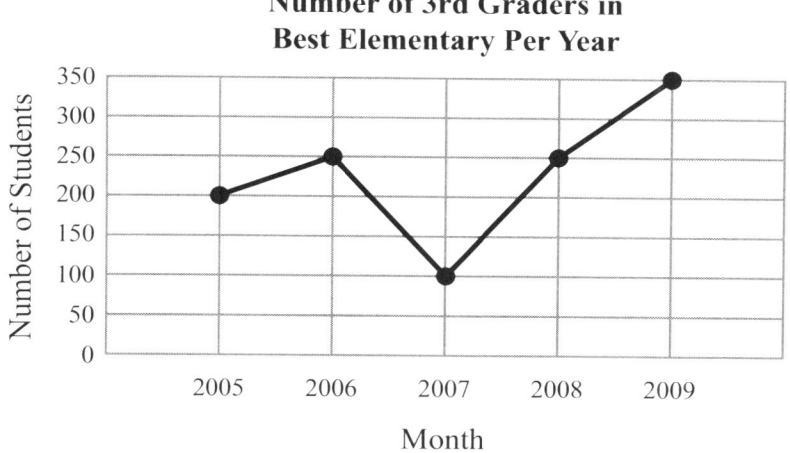

**Number of 3rd Graders in
Best Elementary Per Year**

5. How many 3rd graders were there in 2006? _____

6. Which year had 350 third graders? _____

7. How many more 3rd graders were there in 2009 than 2005? _____

8. Which two years had 250 3rd graders? _____

Use the data in the chart to make a line graph of Jeremy's savings. (DOK 3)

9. Jeremy saved $10.00, $5.00, $20.00, $10.00, and $15.00 each month from January to May. Make a line graph to show how much money Jeremy saved each month. Be sure to label the graph.

Jeremy's Savings	
Month	**Amount**
Jan	$10.00
Feb	$5.00
Mar	$20.00
Apr	$10.00
May	$15.00

7.3 Pictographs (DOK 2)

Pictographs use pictures or symbols to describe a data set. A key is sometimes given to show the value of each picture or symbol. Using a data set, you can create a table to show the frequency of each category.

Example 1: The number of drinks served at a party was recorded. There were 30 colas, 20 diet colas, 10 orange sodas, and 5 lime sodas served. Make a pictograph that describes the data.

Drinks Served	
Type	**Number**
Cola	30
Diet Cola	20
Orange Soda	10
Lime Soda	5

Step 1: Since each number of drinks served is a multiple of 5, let 1 can = 5 drinks.
There were 30 colas served: $30 \div 5 = 6$ cans
There were 20 diet colas served: $20 \div 5 = 4$ cans
There were 10 orange sodas served: $10 \div 5 = 2$ cans
There were 5 lime sodas served: $5 \div 5 = 1$ can

Step 2: Create a table to show the information found in Step 1.

Drinks Served	
Cola	⬭⬭⬭⬭⬭⬭
Diet Cola	⬭⬭⬭⬭
Orange Soda	⬭⬭
Lime Soda	⬭

Key: 1 can = 5 drinks

Example 2: Use the pictograph to find out how many children like banana smoothies.

Smoothies We Like					
Banana	🍌	🍌	🍌		
Grape	🍇	🍇			
Strawberry	🍓	🍓	🍓	🍓	

Key: Each picture = 2 children

Step 1: Look at the pictograph. Find the row with bananas, and count how many bananas are listed.
There are 3 bananas listed.

Step 2: Look at the key. How many children does 1 picture describe?
Each picture describes 2 children.

Step 3: How many children in all like banana smoothies?
$3 \times 2 = 6$

Answer: There are 6 children in all that like banana smoothies.

Use the pictographs to answer the questions. (DOK 2)

Lunches We Like					
Hamburger	🍔	🍔	🍔		
Pizza	🍕	🍕			
Tacos	🌮	🌮	🌮	🌮	

Key: Each Picture = 3 children

1. Which lunch is liked the most?

2. How many children like hamburgers the most?

3. How many fewer children like pizza than tacos?

4. How many children in total like pizza and tacos the most?

Weather for June

Sunny Days	☀	☀	☀	☀	☀	☀	☀
Cloudy Days	☁	☁	☁	☁	☁		
Stormy Days	⚡	⚡	⚡				

Each picture = 2 days

5. How many stormy days were there in June?

6. How many sunny days were there in June?

7. How many more sunny days were there than cloudy days in June?

8. How many days were cloudy and stormy in all?

Use the data in the table to make a pictograph. (DOK 2)

9. A survey of 120 people was conducted to find out what type of movies they like. The survey found that 40 like cartoon movies, 30 like animal movies, 30 like funny movies, and 20 like musical movies. Draw a pictograph using the blank table. The key has been given to you. (DOK 3)

Table:

Type of Movie	Number of People
Cartoon	40
Animal	30
Funny	30
Musical	20

Pictograph:

Type of Movie	Number of People
Cartoon	
Animal	
Funny	
Musical	

Key: 1 circle = 10 people

7.4 Applying Data (DOK 2, 3)

In this section, you will learn to **match data** from one kind of graph to another kind of graph.

Example 1: Decide if the data on the line graph and bar graph match.

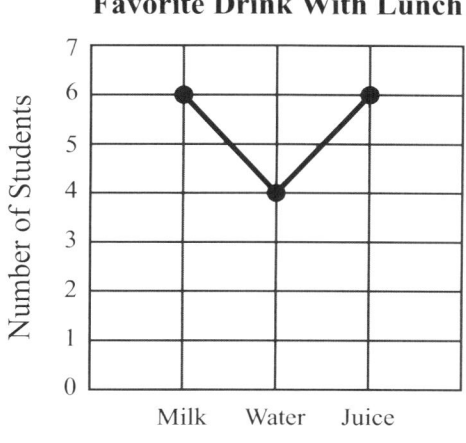

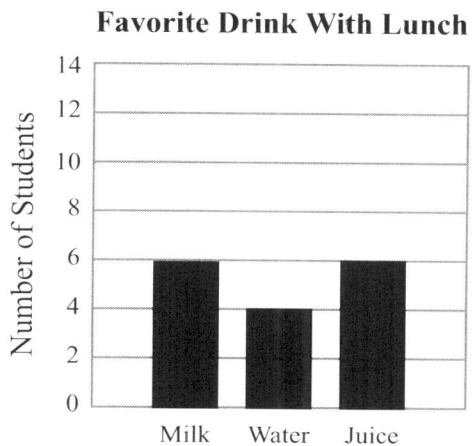

Step 1: Compare the frequencies for the category "Milk." The line graph shows 6 students favor milk. The bar graph shows 6 students favor milk.

Step 2: Compare the frequencies for the category "Water." The line graph shows 4 students favor water. The bar graph shows 4 students favor water.

Step 3: Compare the frequencies for the category "Juice." The line graph shows 6 students favor juice. The bar graph shows 6 students favor juice.

Answer: Since frequencies for each category match, the line graph and the bar graph also match.

1. Which pictograph matches the bar graph below? (DOK 2)

(A)

Number of Butterflies Caught	
Jolie	🦋 🦋 🦋
Jeremy	🦋 🦋 🦋 🦋
Todd	🦋 🦋 🦋

Key: 🦋 = 2 butterflies

(B)

Number of Butterflies Caught	
Jolie	🦋 🦋 🦋 🦋
Jeremy	🦋 🦋 🦋
Todd	🦋 🦋 🦋

Key: 🦋 = 2 butterflies

(C)

Number of Butterflies Caught	
Jolie	🦋 🦋 🦋 🦋
Jeremy	🦋 🦋
Todd	🦋 🦋 🦋

Key: 🦋 = 2 butterflies

2. The Cupcake Bakery keeps track of how many cupcakes they make each day. The pictograph below shows how many cupcakes they made this week. (DOK 2)

Day of the Week	Dozens of Cupcakes
Monday	🧁 🧁 🧁 🧁
Tuesday	🧁 🧁
Wednesday	🧁 🧁 🧁
Thursday	🧁 🧁
Friday	🧁 🧁 🧁 🧁
Saturday	🧁 🧁 🧁 🧁 🧁

Each cupcake = 4 dozen cupcakes

Which bar graph shows the same information as the pictograph?

A)

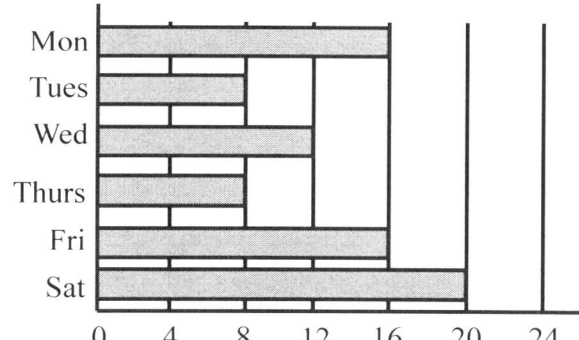

B)

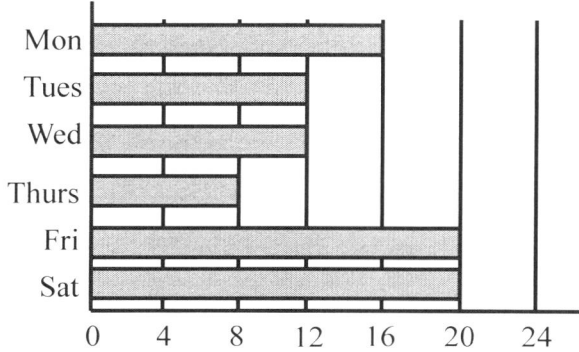

3.   Does the line graph match the bar graph? Explain.

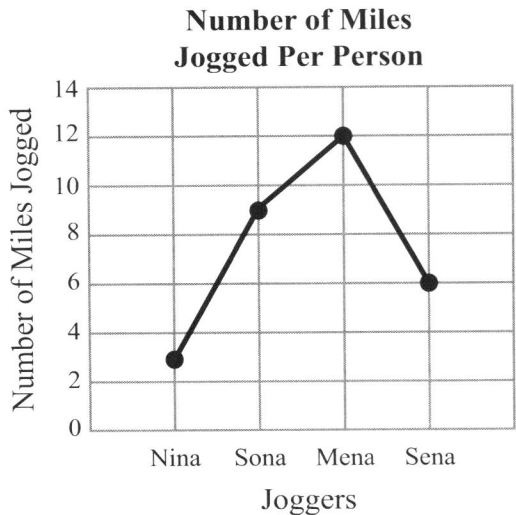
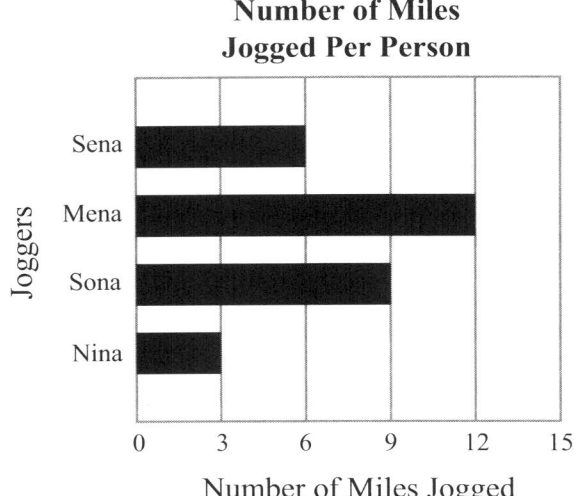

_____

_____

_____

4.   Does the bar graph match the pictograph? Explain.

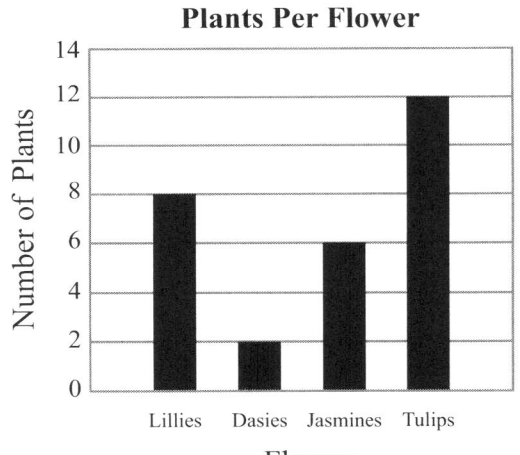

Flower	
Lillies	🌸 🌸 🌸
Daisies	🌸
Jasmines	🌸 🌸
Tulips	🌸 🌸 🌸 🌸 🌸 🌸

**Number of
Plants Per Flower**

Key: 1 flower = 2 plants

_____

_____

_____

Chapter 7 Review

Use the bar graph to answer the questions. (DOK 2)

1. How many games did the Cougars win?

   _____

2. How many more games did the Gators win than the Cougars?

   _____

3. How many games did the Bears and the Eagles win altogether?

   _____

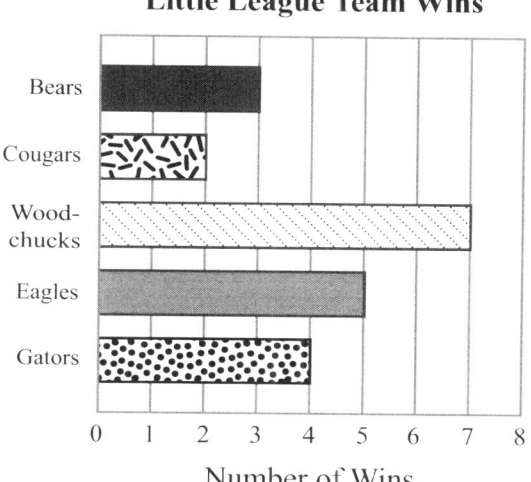

Little League Team Wins

Number of Wins

Use the line graph to answer the questions. (DOK 2)

4. How much money did Emmanuel save in April?

   _____

5. How much more money did Emmanuel save in April than in June?

   _____

6. What does the scale on the vertical side of the line graph measure?

   _____

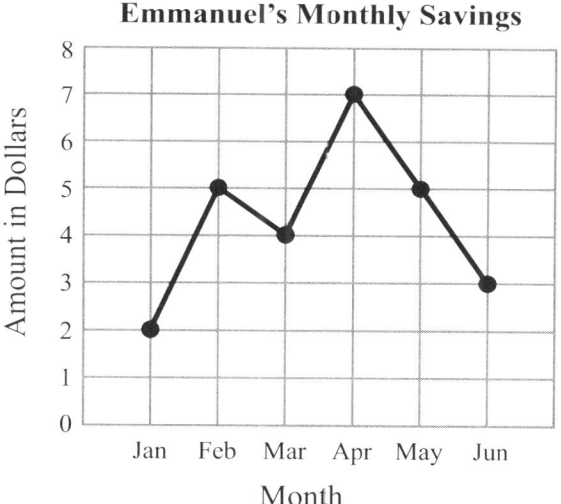

Emmanuel's Monthly Savings

Amount in Dollars

Month

Use the pictograph to answer the questions. (DOK 2, 3)

Number of Each Kind of Tree at School

Number of Each Kind of Tree at School

Oak	🌰	🌰	🌰	🌰	🌰
Elm	🌳	🌳			
Maple	🍁	🍁	🍁	🍁	

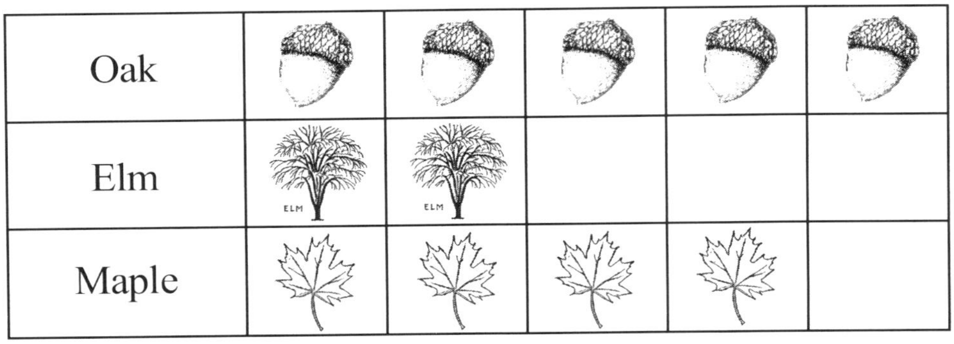

Key: One image = 3 trees

7. How many oak trees are there?

   _____

8. How many more maple trees are there than elm trees?

   _____

9. Complete the table using the information in the pictograph above.

Type of Tree	Number of Trees
Oak	
Elm	
Maple	

10. How many fewer elm trees are there than oak trees?

   _____

11. Which pictograph, A or B, shows the same information as the bar graph? (DOK 2)

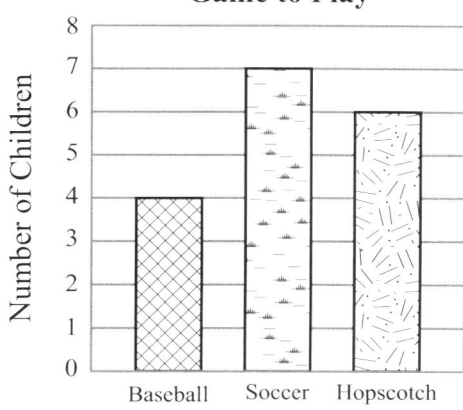

(A)

Game	Number of Children
Baseball	★ ★ ★ ★ ★
Soccer	★ ★ ★ ★ ★ ★ ★
Hopscotch	★ ★ ★ ★ ★ ★
★ = One child	

(B)

Game	Number of Children
Baseball	★ ★ ★ ★
Soccer	★ ★ ★ ★ ★ ★ ★
Hopscotch	★ ★ ★ ★ ★ ★
★ = One child	

12. Create a table using the data in the line graph below. (DOK 3)

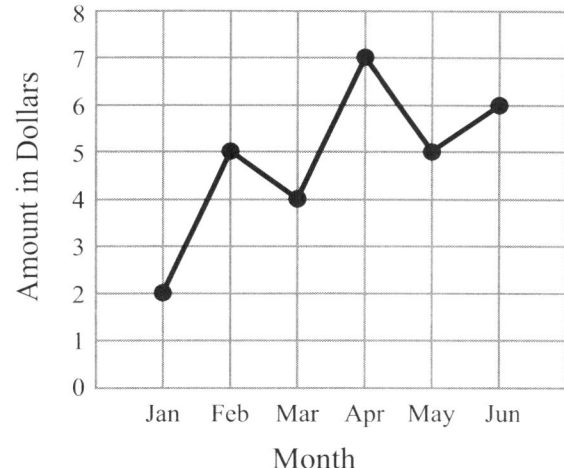

Month	Amount Saved in Dollars
January	
February	
March	
April	
May	
June	

13. Create a table using the data in the line graph below. (DOK 3)

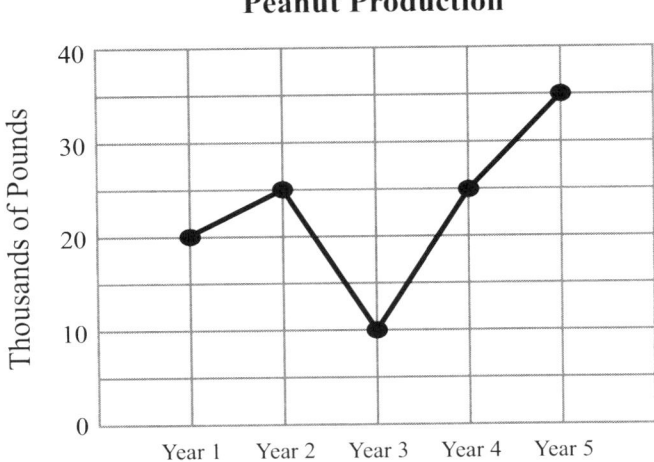

Peanut Production

Year	Thousands of Pounds
Year 1	
Year 2	
Year 3	
Year 4	
Year 5	

14. The bar graph below shows how many kids play a certain sport. Which of the line graphs matches the bar graph below? (DOK 2)

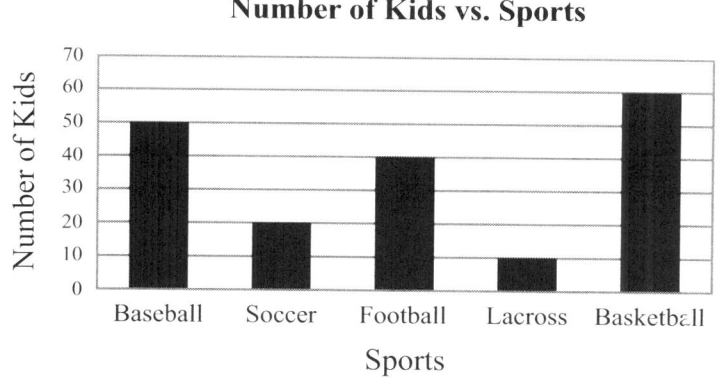

A)

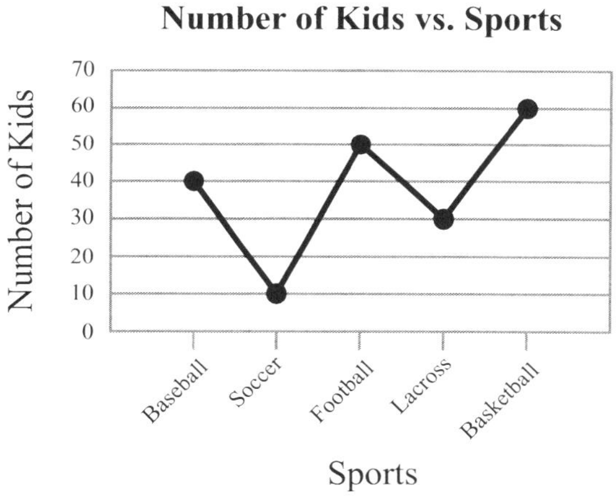

Number of Kids vs. Sports

C)

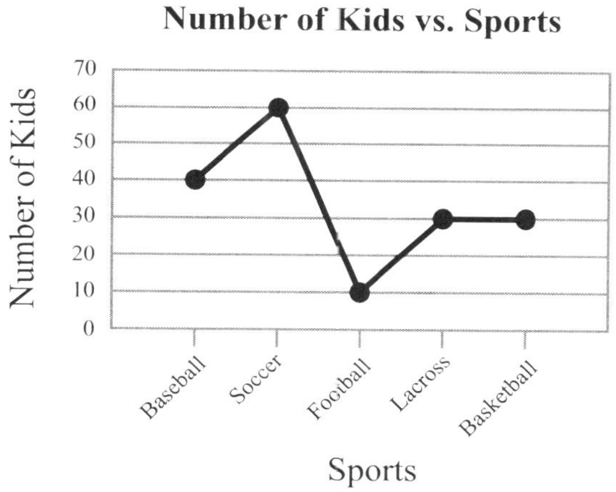

Number of Kids vs. Sports

B)

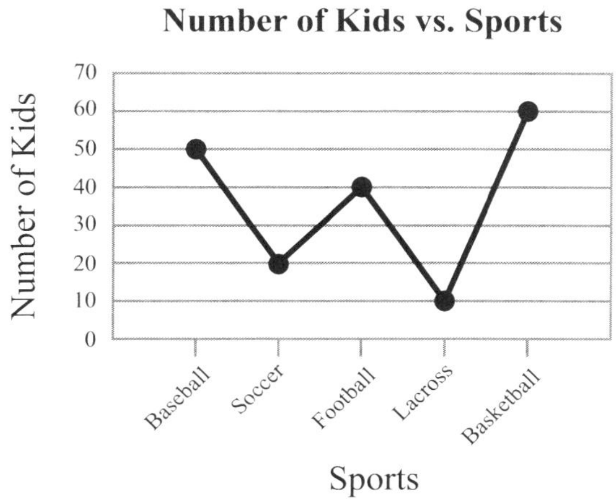

Number of Kids vs. Sports

D)

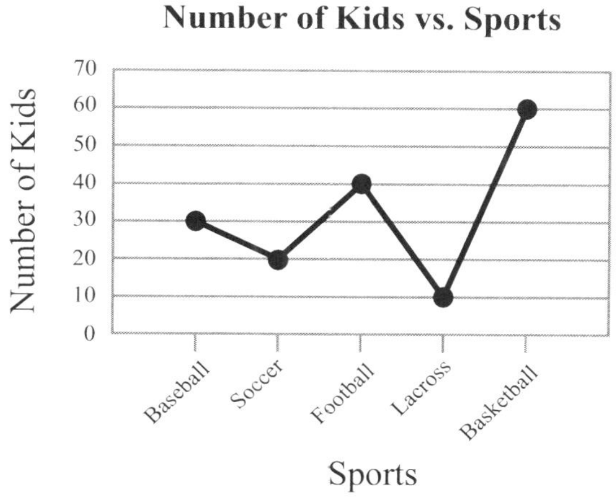

Number of Kids vs. Sports

15. Do the graphs match the data in the table? Explain why on the lines provided below the graphs.

Kids	Presents
Jason	16
John	4
Jacob	20
Joseph	12

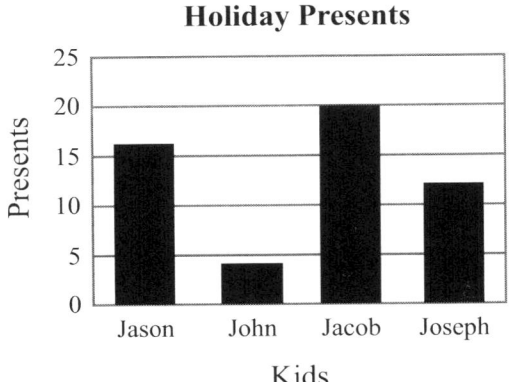

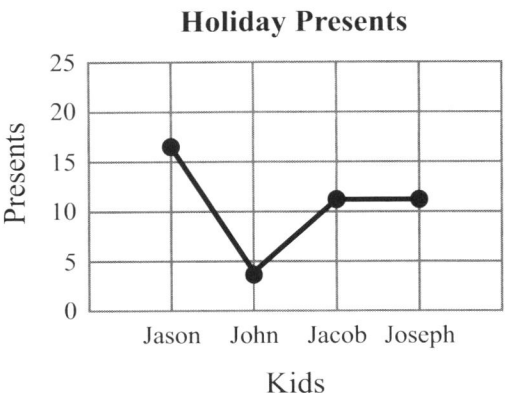

Key: 1 gift box = 4 presents

Kids	Presents
Jason	🎁 🎁 🎁 🎁
John	🎁
Jacob	🎁 🎁 🎁 🎁
Joseph	🎁 🎁 🎁

Bar graph: _____

Line Graph: _____

Pictograph: _____

For additional practice, please see Chapter 7 Test located in the Teacher Guide.

Chapter 8
Geometry

This chapter covers the following Grade 3 standards:

	Content Standard
Measurement and Data	3.MD.5, 3.MD.6, 3.MD.7, 3.MD.8
Geometry	3.G.1, 3.G.2

8.1 Polygons (DOK 1)

A **polygon** is a two-dimensional figure. It has 3 or more sides. Each side meets at a corner to make a closed shape. Polygons can be grouped by the total number of sides and angles they have.

Two-dimensional figures are flat.

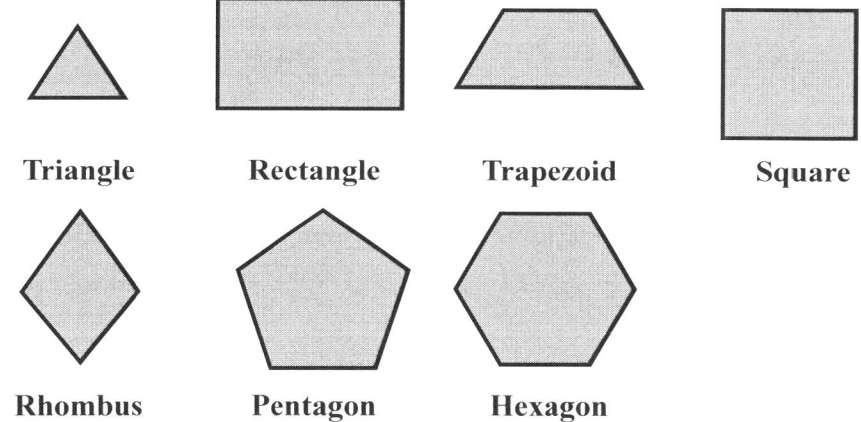

| Triangle | Rectangle | Trapezoid | Square |

| Rhombus | Pentagon | Hexagon |

Below is a chart of the major two-dimensional figures.

Figure	Sides	Angles
Triangle	3 sides	3 angles
Square	4 equal sides	4 equal angles
Rectangle	4 sides; opposite sides are equal.	4 equal angles
Trapezoid	4 sides; sides can have different lengths	4 angles
Rhombus	4 equal sides	4 angles; 2 pairs of equal angles
Pentagon	5 sides; 5 sides	5 angles; a regular pentagon has 5 equal angles
Hexagon	6 sides; 6 sides	6 angles; a regular hexagon has 6 equal angles

Compare the pairs of two-dimensional objects below by giving the name, the number of sides, and the number of angles. (DOK 2)

1.

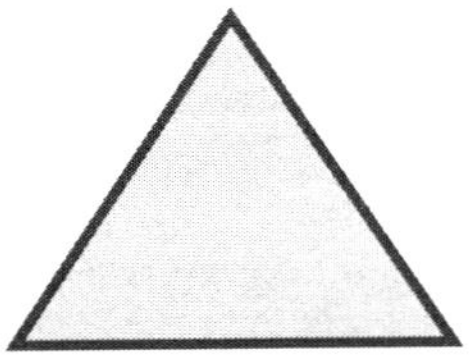

versus

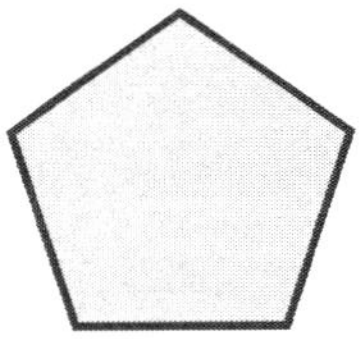

Name: _____

Number of sides: _____

Number of angles: _____

Name: _____

Number of sides: _____

Number of angles: _____

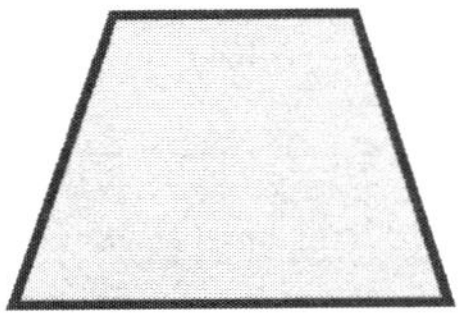

2.

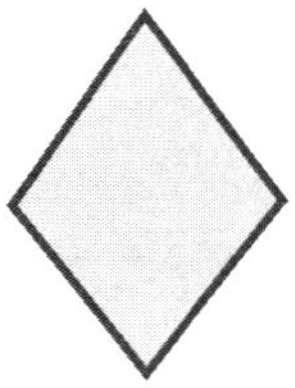

versus

Name: _____

Number of sides: _____

Number of angles: _____

Name: _____

Number of sides: _____

Number of angles: _____

3.

versus

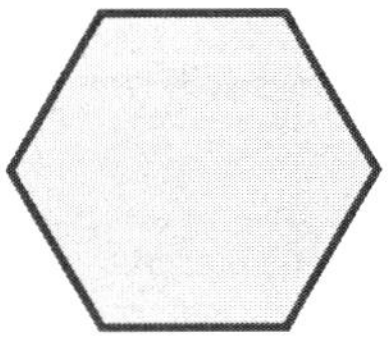

Name: _____

Number of sides: _____

Number of angles: _____

Name: _____

Number of sides: _____

Number of angles: _____

4.

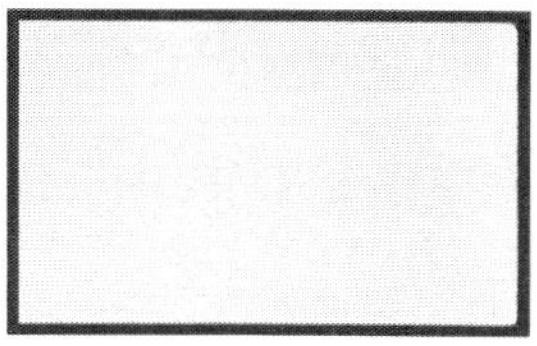

versus

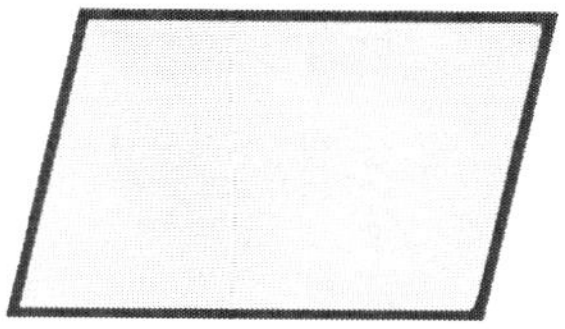

Name: _____

Number of sides: _____

Number of angles: _____

Name: _____

Number of sides: _____

Number of angles: _____

5.

versus

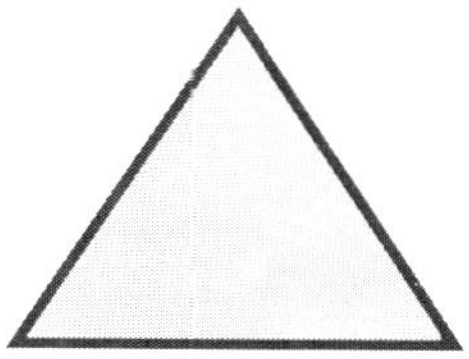

Name: _____

Number of sides: _____

Number of angles: _____

Name: _____

Number of sides: _____

Number of angles: _____

6.

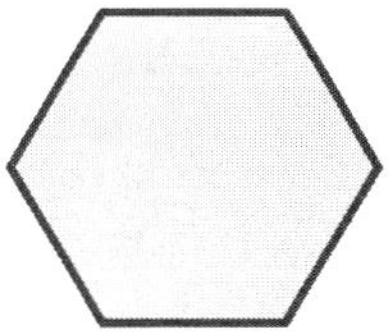

versus

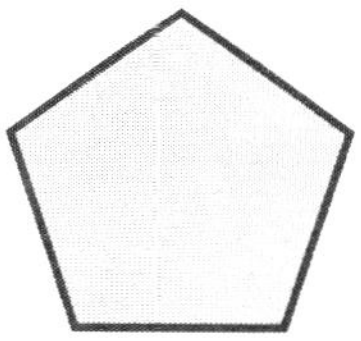

Name: _____

Number of sides: _____

Number of angles: _____

Name: _____

Number of sides: _____

Number of angles: _____

8.2 Area of Rectangles and Squares (DOK 2)

The **area** is the amount of space inside a figure. Area can be found by counting the number of tiles in a given figure or by using multiplication. The unit of measure for area is square units.

The equation for the area of a square or rectangle is length times width.

$$A = length \times width$$

Draw rectangles and squares using tiles. If you cut 6 squares out of paper, you can arrange them into a rectangle. Each square is a square unit. The rectangle below has an area of 6 square units because there are 6 square tiles.

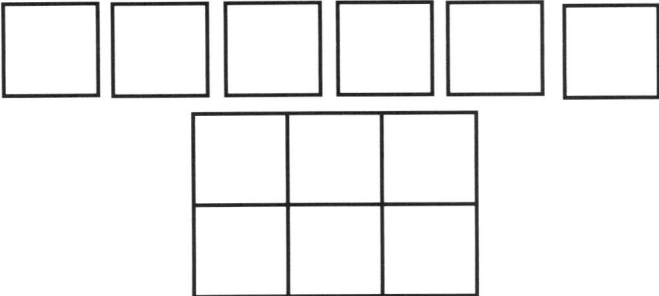

If you cut 4 squares out of paper, you can arrange them into a square. The square below has an area of 4 square units because there are 4 square tiles.

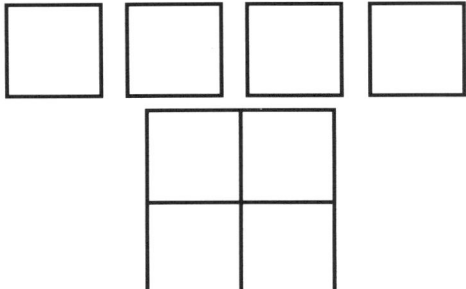

Example 1: Using the graph below, draw a border around 4 columns and 3 rows.

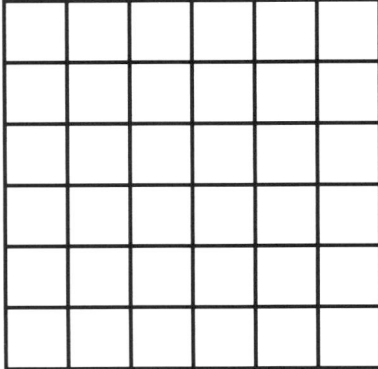

This makes a rectangle.

Count the number of square tiles inside the rectangle.

	1	2	3	4	
	5	6	7	8	
	9	10	11	12	

There are 12 square tiles inside the rectangle. The area of the rectangle is 12 square units.

Example 2: Find the area of the rectangle below.

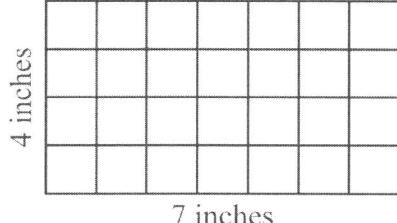

4 inches

7 inches

Step 1: Find the length and width.
Length = 7 and width = 4.

Step 2: Multiply the length and width together.
$7 \times 4 = 28$

Answer: The area of the rectangle is 28 square inches.

Example 3: Find the area of a square that measures 2 cm by 2 cm.

Step 1: The length is 2 cm. The width is 2 cm.

Step 2: Multiply the length and width together.
$2 \times 2 = 4$

Answer: The area of the square is 4 square cm.

Follow the directions for each problem below to find the area of the figure. (DOK 3)

1. Draw a square with 3 rows and 3 columns.

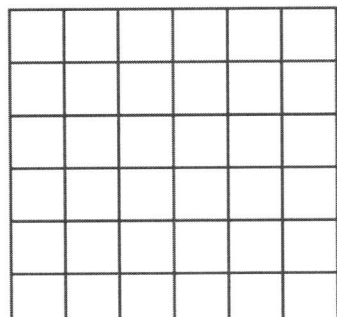

Area = ☐ square units

2. Draw a rectangle with 1 row and 2 columns.

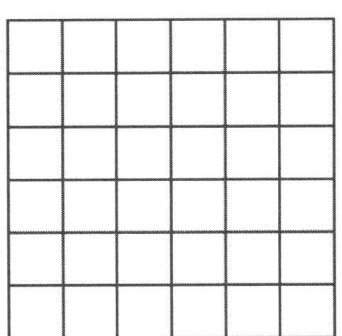

Area = ☐ square units

3. Draw a rectangle with 2 rows and 5 columns.

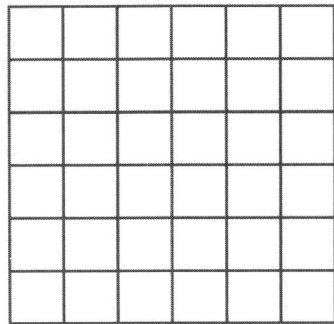

Area = ☐ square units

4. Draw a square with 4 rows and 4 columns.

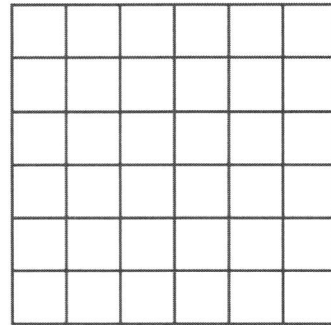

Area = ☐ square units

5. Draw a rectangle with 3 rows and 1 column.

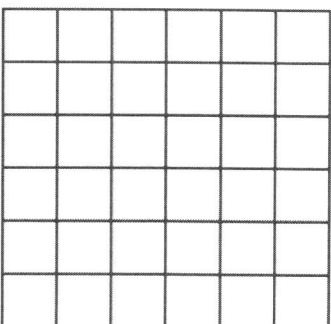

Area = ☐ square units

6. Draw a square with 5 rows and 5 columns.

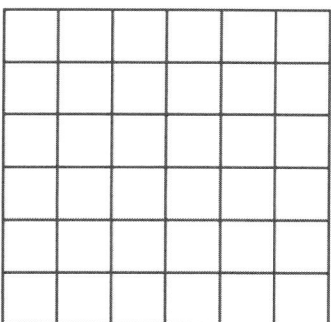

Area = ☐ square units

7.

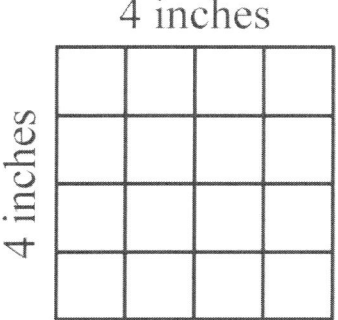

Area = ☐ square units

8.

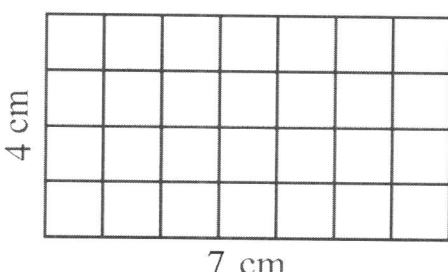

Area = ☐ square units

9.

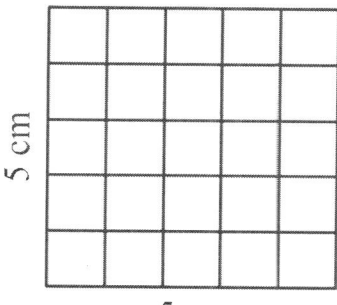

Area = ☐ square units

10.

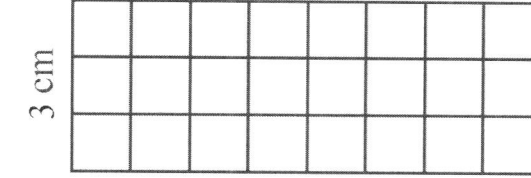

Area = ☐ square units

11.

Area = ☐ square units

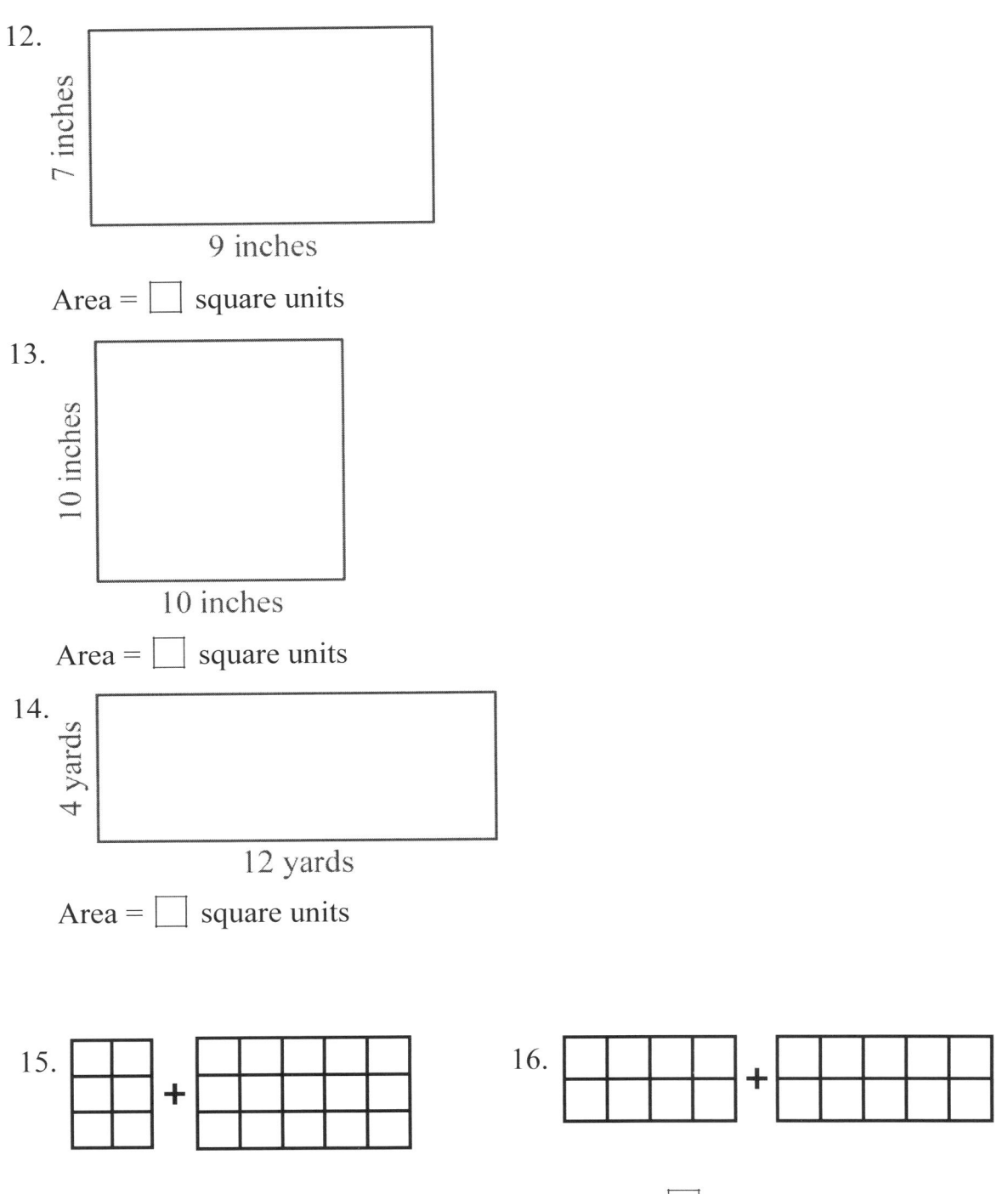

12.

7 inches

9 inches

Area = ☐ square units

13.

10 inches

10 inches

Area = ☐ square units

14.

4 yards

12 yards

Area = ☐ square units

15. ☐ + ☐

Area = ☐ square units

16. ☐ + ☐

Area = ☐ square units

8.3 Perimeter (DOK 1, 2)

Perimeter is the distance around the outside of an object. For example, the distance around the outer edge of a piece of paper is the perimeter. The perimeter is found by adding the side lengths of a figure.

Example 1: Find the perimeter of the trapezoid below.

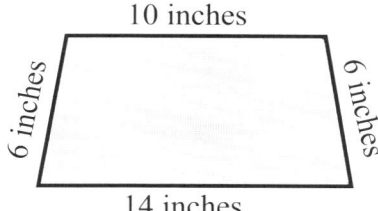

Step 1: List the side lengths.
6, 10, 6, 14

Step 2: Add the side lengths.
$6 + 10 + 6 + 14 = 36$

Answer: The perimeter is 36 inches.

Remember to always include the unit of measure.

Example 2: The triangle has a perimeter of 28 inches. Find the missing side length.

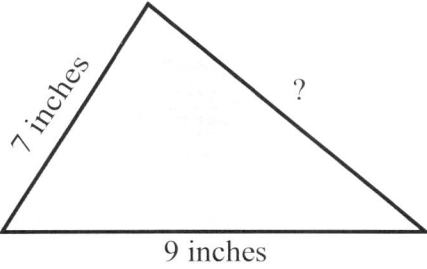

Step 1: Add the side lengths of the 2 sides that are given.
$7 + 9 = 16$

Step 2: Subtract the value found in step 1 from the perimeter.
$28 - 16 = 12$

Answer: The missing side length is 12 inches.

Example 3: The triangle has a perimeter of 12 inches. Find the measure of the missing side

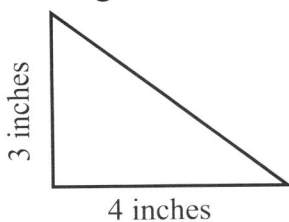

3 inches

4 inches

Step 1: Add the side lengths of the 2 sides that are given.
3 + 4 = 7

Step 2: Subtract the value found in step 1 from the perimeter.
12 − 7 = 5

Answer: The missing side length is 5 inches.

Find the perimeter. (DOK 1)

1. Perimeter = ☐ inches

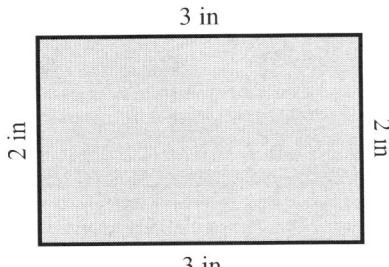

3 in

2 in   2 in

3 in

3. Perimeter = ☐ cm

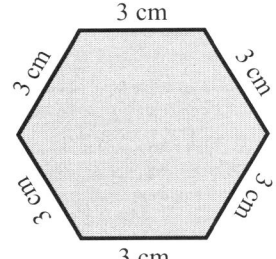

3 cm

3 cm   3 cm

3 cm   3 cm

3 cm

2. Perimeter = ☐ inches

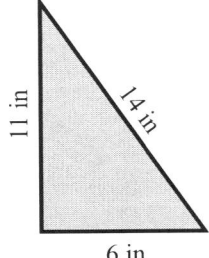

11 in   14 in

6 in

4. Perimeter = ☐ cm

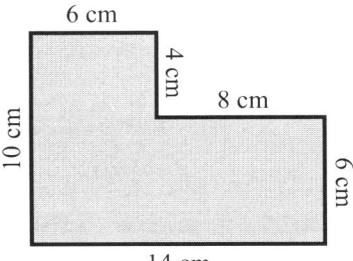

6 cm

4 cm

8 cm

10 cm   6 cm

14 cm

5. Perimeter = ☐ feet

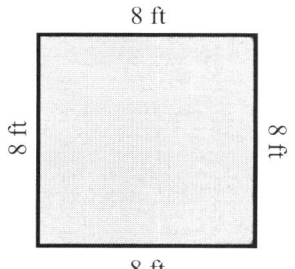

9. Perimeter = ☐ ft

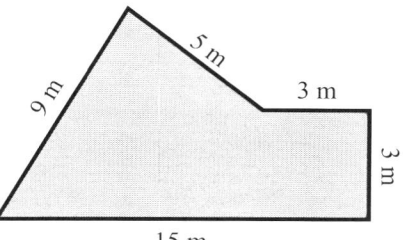

6. Perimeter = ☐ m

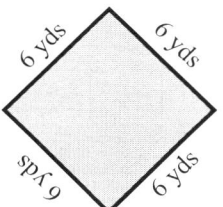

10. Perimeter = ☐ meters

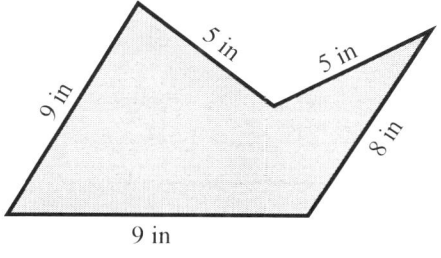

7. Perimeter = ☐ yards

8. Perimeter = ☐ inches

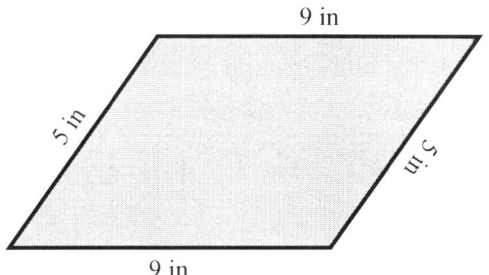

11. Perimeter = ☐ ft

12. Perimeter = ☐ inches

Find the missing side length in each figure below. Be sure to include the unit of measure. (DOK 2)

13. Perimeter = 38 cm

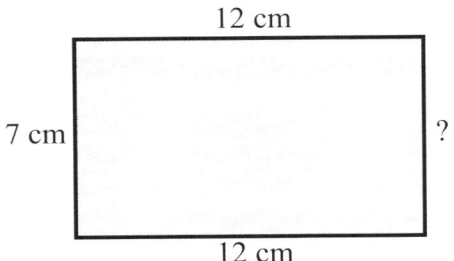

Missing side length = _____

14. Perimeter = 20 ft

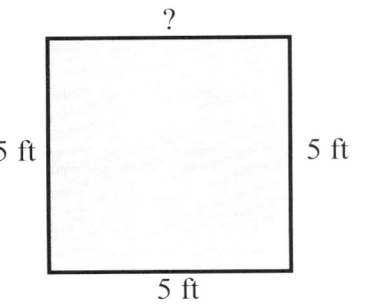

Missing side length = _____

15. Perimeter = 28 inches

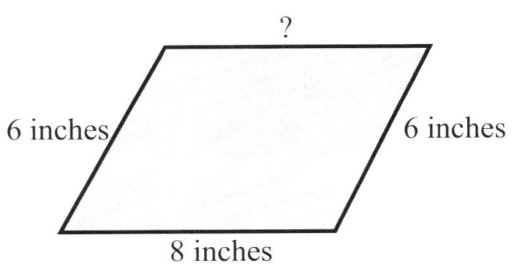

Missing side length =_____

16. Perimeter = 40 in

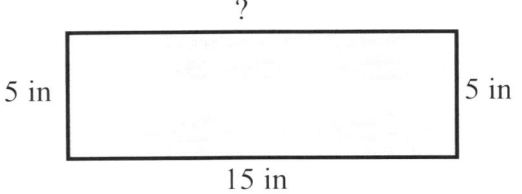

Missing side length = _____

17. Perimeter = 50 inches

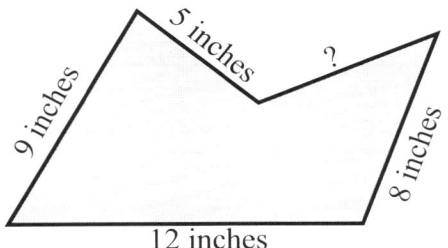

Missing side length = _____

18. Perimeter = 52 cm

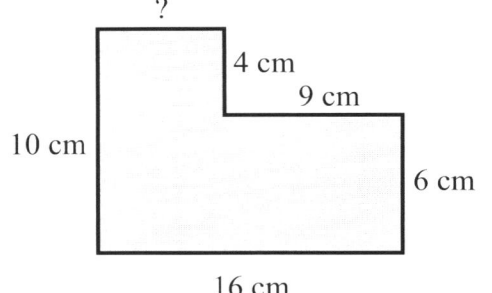

Missing side length = _____

19. Perimeter = 60 in

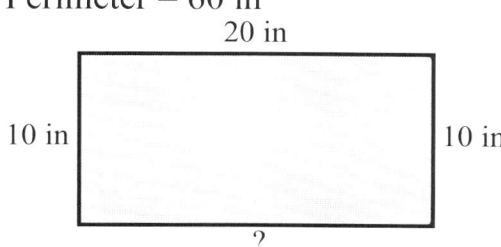

Missing side length = _____

20. Perimeter = 80 ft

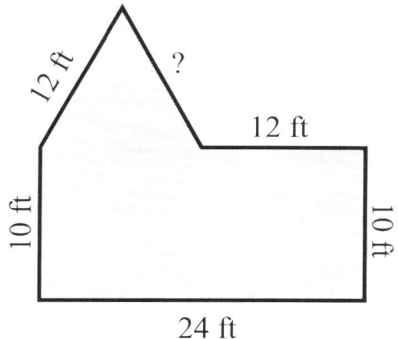

Missing side length = _____

21. Perimeter = 44 cm

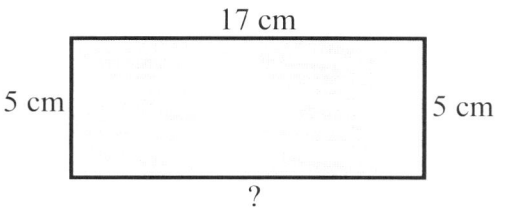

Missing side length = _____

22. Perimeter = 32 in

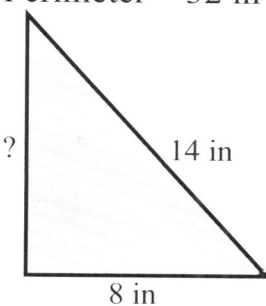

Missing side length = _____

23. Perimeter = 60 cm

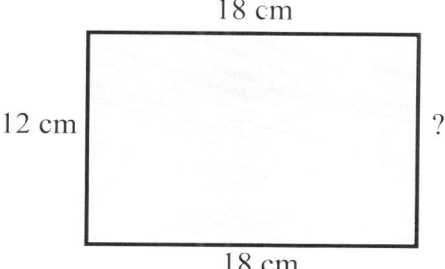

Missing side length = _____

24. Perimeter = 35 in

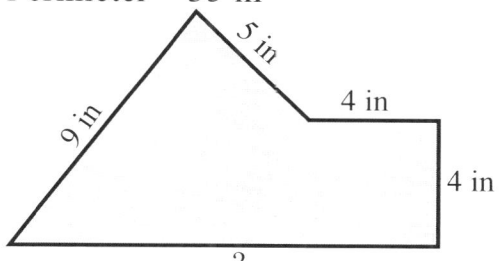

Missing side length = _____

8.4 Using Perimeter and Area (DOK 2)

Example 1: Mr. Wranger wants to place an advertisement in a newspaper. There are two newspapers in town. He found the prices for 2 differently sized advertisement spaces. Compare the perimeter and area of each advertisement space.

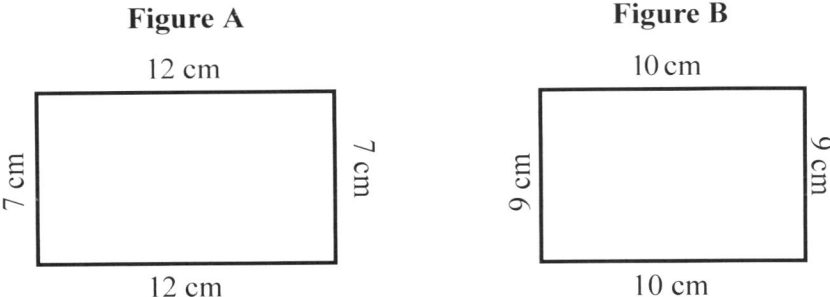

Step 1: Find the perimeters and areas of the 2 advertisement spaces.
Figure A
Perimeter: 7 + 12 + 7 + 12 = 38 cm
Area: 7 × 12 = 84 square cm

Figure B
Perimeter: 9 + 10 + 9 + 10 = 38 cm
Area: 9 × 10 = 90 square cm

Step 2: Compare the two advertisement spaces (see conclusion below).

Conclusion: Even though the two ad spaces have different measurements, they both have the same perimeter of 38 cm.
However, they have different areas: Figure A = 84 square cm, and Figure B = 90 square cm.

Find the perimeter and areas of the pairs of figures below. Write a comparison sentence for each pair of figures. (DOK 2)

1.

Figure A   **Grandmother's Gardens**   **Figure B**

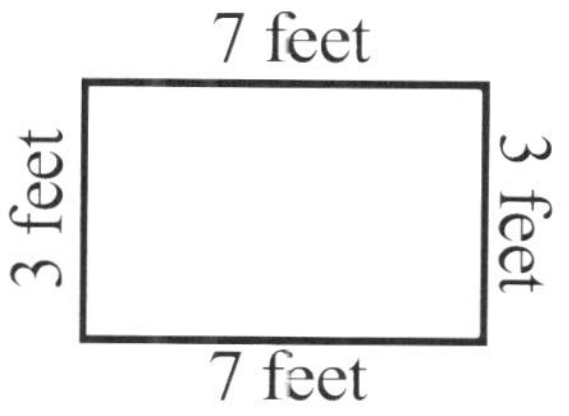

Figure A: square, 5 feet, 5 feet, 5 feet, 5 feet

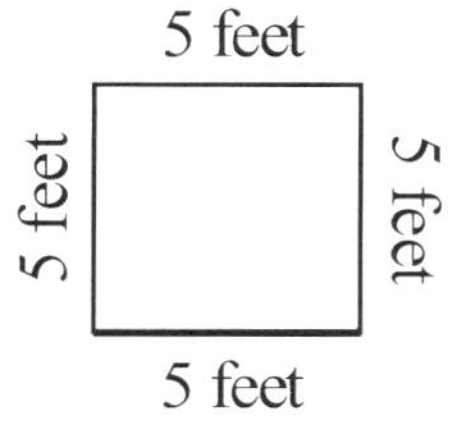

Figure B: rectangle, 7 feet (top), 3 feet (left), 3 feet (right), 7 feet (bottom)

Perimeter: _____   Perimeter: _____

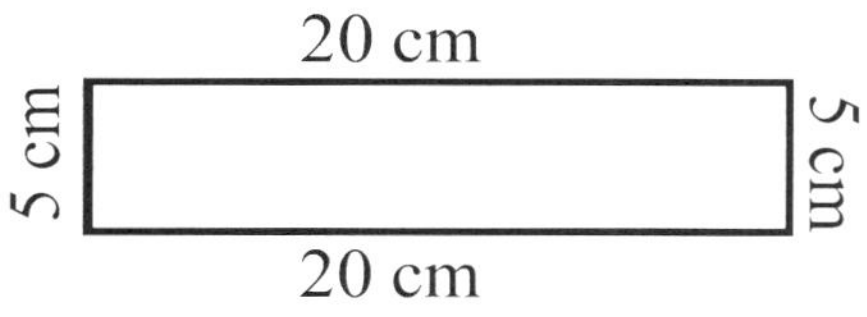

Area: _____   Area: _____

Comparison: _____

2.

Figure C   **Bandages**   **Figure D**

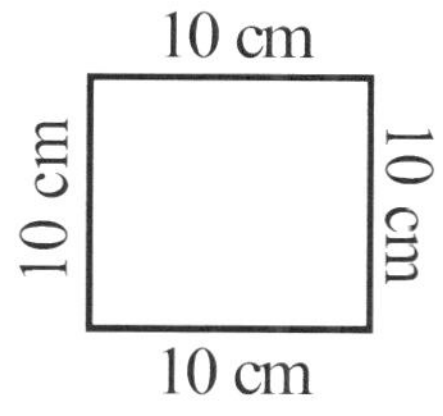

Figure C: rectangle, 20 cm (top), 5 cm (left), 5 cm (right), 20 cm (bottom)

Figure D: square, 10 cm, 10 cm, 10 cm, 10 cm

Perimeter: _____   Perimeter: _____

Area: _____   Area: _____

Comparison: _____

3.

Figure E   **Pieces of Paper**   **Figure F**

A piece of paper that measures 7 inches by 8 inches.   A piece of paper that measures 9 inches by 6 inches.

Perimeter: _____   Perimeter: _____

Area: _____   Area: _____

Comparison: _____

8.5 Applying Area (DOK 2, 3)

Example 1: Mr. Finigan has a piece of stainless steel that measures 9 cm wide and 6 cm long. He wants to divide the piece of stainless steel into 3 equal parts.

Step 1: Mr. Finigan finds the area of the piece of stainless steel: $9 \times 6 = 54$ square cm.

Step 2: He divides the area by 3 to get one third of a piece: $54 \div 3 = 18$ square cm.

Answer: Did Mr. Finigan divide the piece of stainless steel correctly? To find out, take the measures of one piece and find the area; $3 \times 6 = 18$ square cm. Yes, he divided the piece of stainless steel into 3 equal parts.

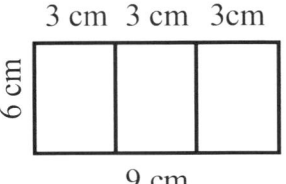

Example 2: Mrs. Thompson is laying down square blocks to make a patio and wants to know the final measurements of the patio. Each square block measures 10 inches wide and 10 inches long. She wants the patio to measure 12 blocks wide and 18 blocks long. What will be the final measurements of the patio?

Step 1: Mrs. Thompson takes the width, 12 blocks, and multiplies by the measure of the square patio block, 10 inches; $121 \times 10 = 120$ inches wide.

Step 2: Mrs. Thompson takes the length, 18 blocks, and multiplies by the measure of the square patio block, 10 inches; $18 \times 10 = 180$ inches long.

Answer: The final measurements of the patio are 120 inches wide by 180 inches long.

Carefully read each problem and solve. (DOK 2, 3)

1. Nathan has a piece of plywood that measures 8 feet long and 4 feet wide. He wants to cut the piece of plywood into 4 equal pieces. What is the area of $\frac{1}{4}$ of a piece of plywood?

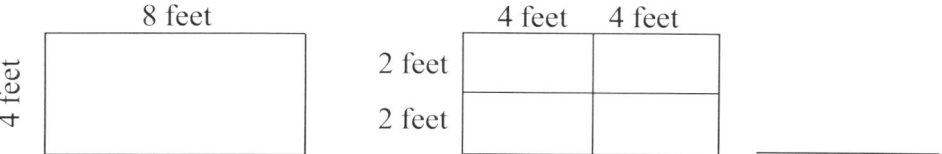

2. Myra has 3 pieces of felt. Each piece measures 5 inches wide by 8 inches long. She wants to glue them to a piece of cardboard, like the drawing to the right. What is the area of the final project?

_____

3. Mr. Tolefson is tiling the top of a coffee table. Each tile is 6 inches long and 6 inches wide. The top of the table measures 30 inches wide by 48 inches long. How many tiles will Mr. Tolefson need to cover the table? To figure this out, divide each measure of the table top by the measure of the tile. Then, multiply the number of tiles used wide and long to find the number of tiles needed.

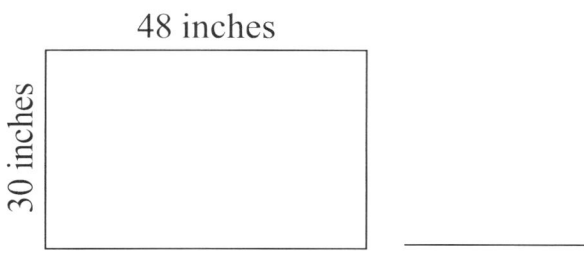

Page 191

4. The picture below shows the size of a piece of wrapping paper Abby has left over.

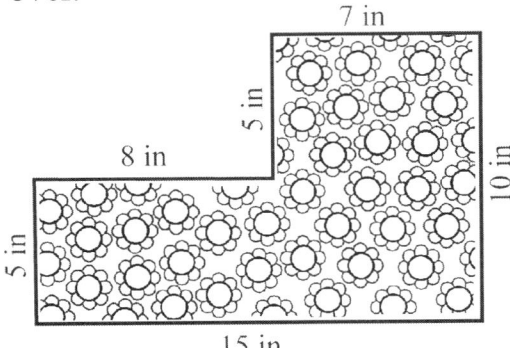

How can the area of the piece of wrapping paper be found?
A) $(5 \times 7) + (5 \times 15)$
B) $(5 + 7) \times (5 + 15)$
C) $(5 \times 7) \times (5 \times 15)$
D) $(5 \times 10) + (5 \times 15)$

5. The picture below shows the outline of Andrew's lizard sticker.

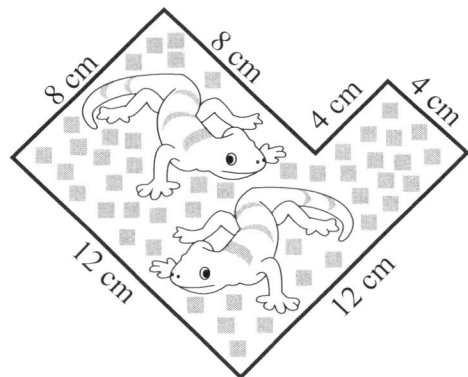

How can the area of the sticker be found?
A) $(8 \times 8) + (4 \times 4)$
B) $(8 + 8) \times (4 + 4)$
C) $(8 \times 12) \times (4 + 4)$
D) $(8 \times 12) + (4 \times 4)$

6. Harold has a piece of cardboard that has the same measurements as the drawing below.

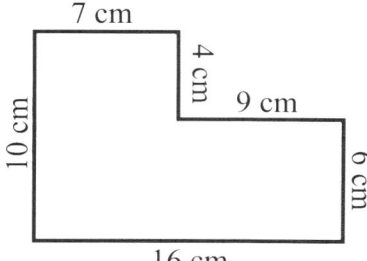

How can the area of the piece of cardboard be found?

A) $(4 + 7) \times (6 + 16)$

B) $(4 \times 7) + (6 \times 10)$

C) $(7 \times 10) + (9 \times 6)$

D) $(7 \times 10) + (9 \times 16)$

7. Derek has a scrap piece of plywood that has the same measurements as the drawing below.

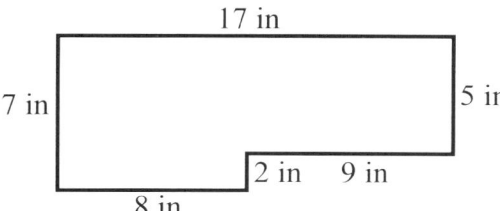

How can the area of the plywood be found?

A) $(5 \times 17) + (2 \times 8)$

B) $(5 + 17) \times (2 + 8)$

C) $(7 \times 8) + (9 \times 2)$

D) $(7 + 8) \times (9 \times 5)$

8.6 Geometry Enrichment (DOK 3)

Carefully read each problem below and solve. Show your work. (DOK 3)

1. Mr. Jackson has a garden that measures 10 feet by 10 feet. He wants to extend one side 5 feet out. What will the perimeter and area of the garden be when Mr. Jackson finishes expanding his garden?

   Perimeter: _____ Area: _____

2. Lisa has a piece of paper measuring 8 cm by 12 cm. She cuts off a 3 cm by 12 cm strip off the edge. What is the perimeter and area of the piece of paper after Lisa trims off one edge?

   Perimeter: _____ Area: _____

3. Mr. Wilcox has a deck on the back of his house that measures 9 feet by 14 feet. He decides to extend it to a new size of 14 feet by 14 feet. He needs to know how many more square feet his deck will be so he can buy the correct amount of wood. Find the area of just the new portion of the deck.

   Area: _____

4. The gym teacher at Golden Valley Elementary, Mr. Ridley, has a path set up outside for the kids to run on. It is shaped like a rectangle and measures 100 feet by 20 feet. He decides to widen the path to a total of 100 feet by 30 feet. What is the measure of the perimeter of the new path?

   Perimeter: _____

Chapter 8 Review

Write the name of the figure on the line. (DOK 2)

1.

_____

2.

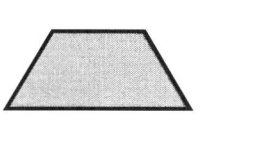

_____

3.

_____

4.

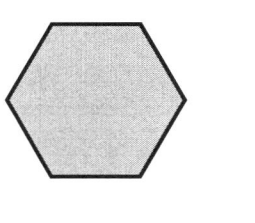

_____

5.

_____

6.

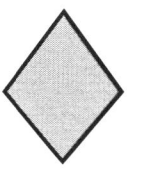

_____

7.

_____

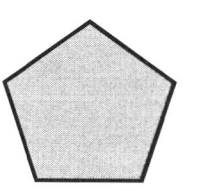

8. List the numbers from questions 1-7 that have four sides.

    _____

9. List all the shapes from questions 1-7 that have more than four sides.

    _____

(DOK 2, 3)

10. Draw a rectangle with 2 rows and 4 columns. Find the area.

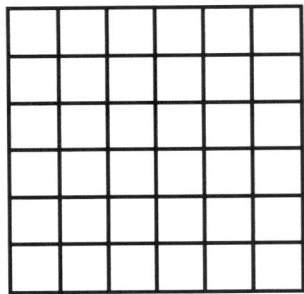

Area = ☐ square units

11. Find the perimeter and area of the figures below. Then, write a comparison sentence about the two figures.

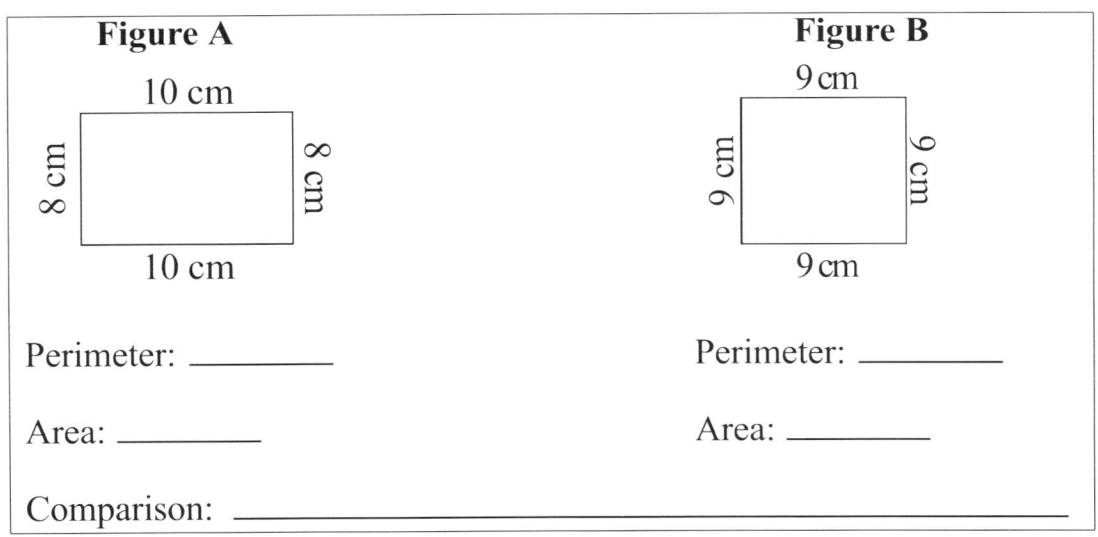

Figure A	Figure B
10 cm (top), 8 cm (left), 8 cm (right), 10 cm (bottom)	9 cm (top), 9 cm (left), 9 cm (right), 9 cm (bottom)
Perimeter: _____	Perimeter: _____
Area: _____	Area: _____
Comparison: _____	

12. Laura has 4 squares of felt. Each square has side lengths of 6 inches. What is the area of each of the squares of felt?

_____

13. Mrs. Bridges' table is 6 feet long and 3 feet wide. She adds an extension to the table that is 2 feet long and 3 feet wide. What is the perimeter and area of the table with the extension?

Perimeter: _____

Area: _____

14. The figure below has a perimeter of 48 cm. What is the missing side length?

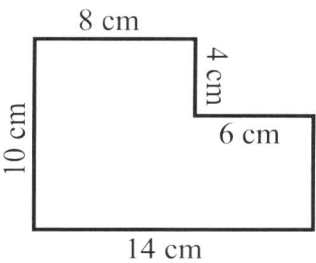

_____

15. Aaron has a piece of cardboard that is 16 inches wide and 20 inches long. He cuts the cardboard into 4 equal pieces. Each smaller piece has a width of 8 inches. What is the new length of each smaller piece?

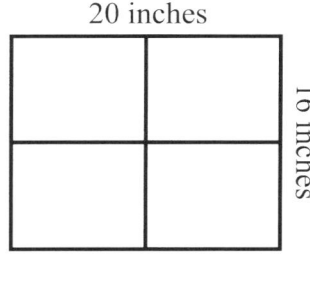

_____

16. Draw a square with 5 rows and 5 columns. Find the area.

Area = ☐ square units

17. Find the total area of the images below.

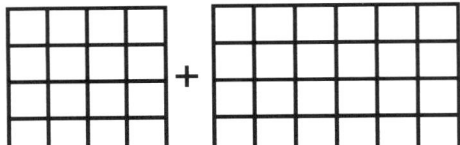

For additional practice, please see Chapter 8 Test located in the Teacher Guide.

Chapter 9
How to Answer Next Generation Assessment Questions

You may be familiar with answering multiple choice questions on tests. These questions ask you to choose an answer from answer A, answer B, answer C, or answer D. But many tests ask different types of questions:

- choose multiple answers
- circle or highlight answers
- fill in the blank
- extended response answers
- click and drop answers (on computerized testing)
- order numbers or objects
- match questions to answers
- draw points on graphs

This chapter will help you get ready to answer those types of questions. When you see a question, read it carefully. Make sure you understand exactly what the question is asking. If you need help, ask your teacher.

9.1 Multiple Choice/Multiple-Multiple Choice

Some Common Core questions ask you to choose the correct answer, or answers, from choices given. You may be given the test on a computer, or it might be printed out on paper for you. The questions you will answer can either be regular multiple choice (print and computer), multiple-multiple choice (print and computer), circle (print), or drop-down multiple choice (computer). Look at the following examples.

Example 1:   Which number is a factor of 64 and a multiple of 4?

A)   3

B)   5

C)   7

D)   8

The correct answer to select would be D) 8.

Example 2:   Which numbers are factors of 64? Select all that are correct.

☐ 2   ☐ 3
☐ 4   ☐ 5
☐ 6   ☐ 7
☐ 8   ☐ 9

The correct answers would be 2, 4, and 8.

Example 3:   Circle all of the shapes below that have two sets of parallel lines.

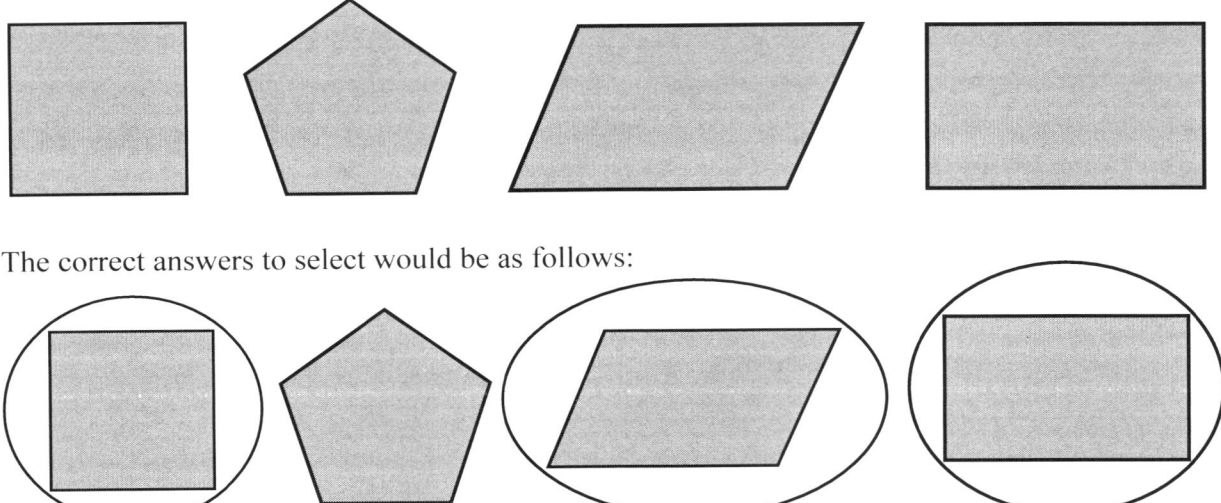

The correct answers to select would be as follows:

Multiple choice questions on the computer might look like these examples below:

Example 4:   **Multiple Choice**

Select the correct answer from the list below the question.

Example: 8 + 4 =

○ 12
○ 18
○ 10
○ 17

Example 5:

Which of the following are factors of 64? Select all that are correct.

✔☑ 2
☐ 3
✔☑ 4
☐ 5
☐ 6
☐ 7
✔☑ 8

Example 6:

Drop Down

Click the drop down box to show a list of choices below. Select the correct answer.

Example: 3 + 6 = | --- |

Drop Down

Click the drop down box to show a list of choices below. Select the correct answer.

Example: 3 + 6 =

10
11
8
9

Drop Down

Click the drop down box to show a list of choices below. Select the correct answer.

Example: 3 + 6 = | 9 |

9.2 Fill in the Blank

Some questions ask you to fill in the blank. This means you will not be given any answer choices; you will have to think of the answer on your own. Whatever the question asks, be sure to answer it clearly. Here is an example.

Tammy needs $50 to go to her school dance. She has 5 family members willing to give her an equal amount of money so that she can go. How much money does each family member need to give Tammy? Write your answer on the line below.

_____$10_____

This is a division problem, so you would need to divide $50 by 5 family members: $50 ÷ 5 = $10.

A fill in the blank question on the computer might look like this example below:

Example 1:

Fill in the Blank

Click inside the white box next to a question to type in an answer.

Example: 2 + 2 =   ⬚

Fill in the Blank

Click inside the white box next to a question to type in an answer.

Example: 2 + 2 =   ⬚ 4

9.3 Extended Response

At times, the test will ask you to do more than just choose correct answers or fill in the blank. Sometimes, you will be asked to write extended response answers to questions. Here is an example. Read the question carefully. Look at exactly what it asks you to do. Then, practice answering the question. You will also see how the best answer looks.

Example 1: Annie runs a soup kitchen on Saturdays in the inner city. In the last few years, as hard times have hit, Annie has noticed a greater use of her soup kitchen. The table below shows the number of people she feeds per week.

Year	1	2	3	4	5
# of people fed (per week)	50	65	80	95	?

A) Describe how to find the next value in the table, using words. Explain how you came to your answer.

B) How many people will use Annie's Soup Kitchen per week in year 5 if the pattern continues?

C) Write a number sentence using the letter n to represent the number of people fed in year 5.

Good Answer:

A) Add 15 to the number of people fed each week. 95

B) 110 people

C) $95 + 15 = N.$

The good answer has correct answers for each part of the question.

Better Answer:

A) Add 15 to the number of people fed each week. $50 + 15 = 65$; $65 + 15 = 80$; $80 + 15 = 95$.

B) $95 + 15 = 110$; 110 people

C) Add 15 to the number of people fed in year 4 to find the number of people fed in year 5. $95 + 15 = n$.

The better answer shows the math used to come up with the correct answer for each part of the question.

Best Answer:

A) Each year the number of people fed each week increases by 15.
   Add 15 to the number of people fed each week to find the number of people fed each week in the next year.
   $50 + 15 = 65$; $65 + 15 = 80$; $80 + 15 = 95$.

B) Add 15 to the number of people fed in year 4 to find the number of people fed in year 5.
   $95 + 15 = 110$.
   If the pattern continues, the number of people fed in year 5 will be 110.

C) Let n represent the number of people fed in year 5.
   Add 15 to the number of people fed in year 4 to find the number of people fed in year 5.
   $95 + 15 = n$.

The best answer uses full sentences to answer each part of the question, shows the work used to come up with correct answers for each part of the question, and has a detailed explanation for why the answer is correct.

An extended response question on the computer might look like this example below:

Example 2:

Extended Response

Click in the box to type in an extended response.

Example:

Each year the number of people fed each week is increasing by 15.

Add 15 to the number of people fed each week to find the number of people fed each week in the next year.

$50 + 15 = 65$; $65 + 15 = 80$; $80 + 15 = 95$

9.4 Click and Drop

Your computerized test may ask you to click and drop selected answers.

A click and drop question on the computer might look like this example below:

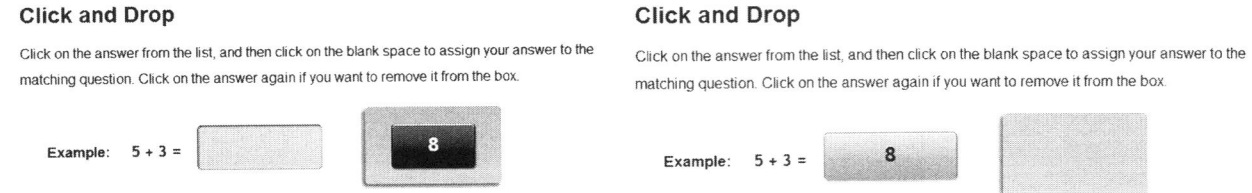

9.5 Ordering

Your computerized test may ask you to click and drop selected answers in the correct order.

An ordering question on the computer might look like this example below:

Ordering

Click on the answer or phrase, and then click on a different answer or phrase to switch their places.

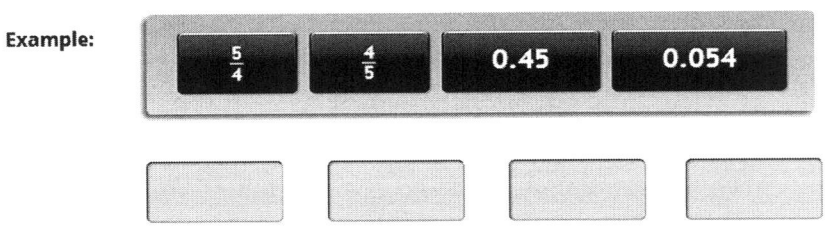

Ordering

Click on the answer or phrase, and then click on a different answer or phrase to switch their places.

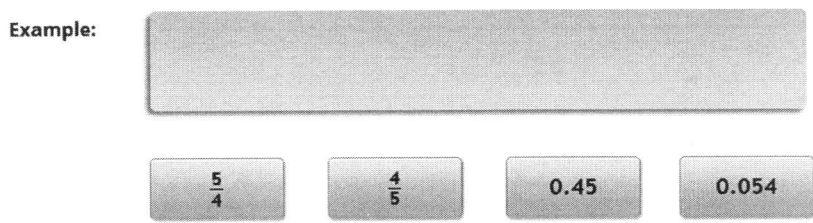

9.6 Matching (Connection)

Your computerized test may ask you to match questions with the correct answers.
A matching question on the computer might look like this example below:

Connection

Click on the question, and then connect it to its answer by clicking on the matching answer.
Repeat the process to remove the connection.

Example:

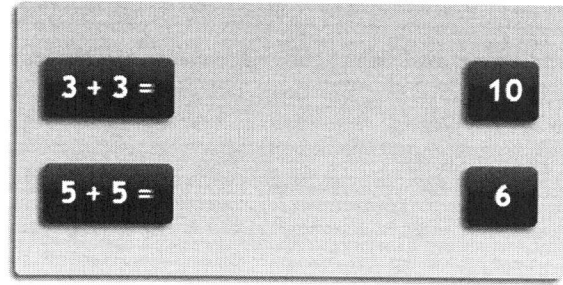

Connection

Click on the question, and then connect it to its answer by clicking on the matching answer.
Repeat the process to remove the connection.

Example:

9.7 Plot Points

Your computerized test may ask you to plot points on a graph.

A point plotting question on the computer might look like this example below:

Plot (points)

Select the correct point(s) on the plot below that correspond to the question.

Example: Select the point (1, 1).

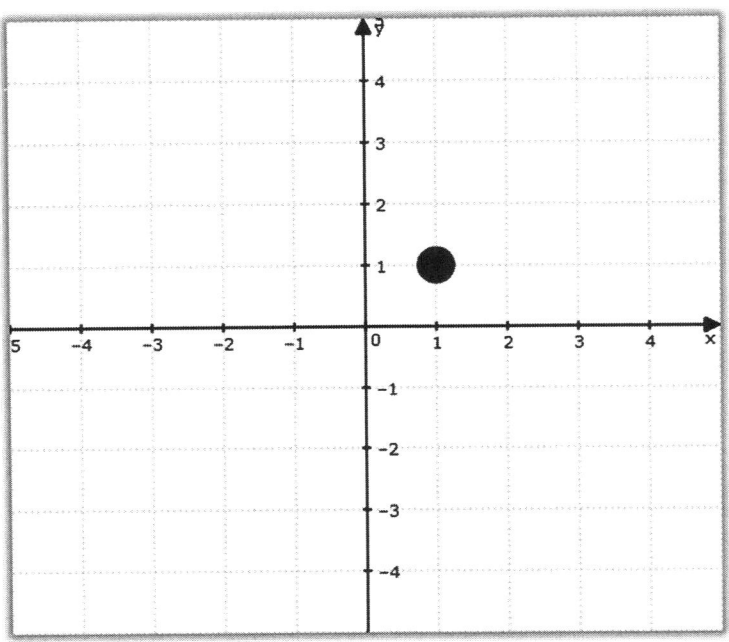

9.8 Highlight Errors

Your computerized test may ask you to highlight errors in a problem.

A highlighting errors question on the computer might look like this example below:

Sergio was told to evaluate the following problem.

$$[8(4+2)] - [3(5-1)]$$

He wrote the following steps.

$$[32 + 16] - [3(5 - 1)]$$

$$[32 + 16] - [15 - 3]$$

$$44 - 12$$

$$32$$

Highlight the step where Sergio made his mistake.

9.9 Performance Tasks

Your test will include a performance task. These tasks will give you a real-life situation to work with. Performance tasks test how well you can combine many concepts and skills you know into a project. You will use sources that are given to you. Then, you will answer six questions about the project that are like the types of questions you saw in this chapter.

Post Test

Follow the directions for each question.

1   Circle true or false for each question.

    A $7 \times 8 = 56 \div 7$ True   False

    B $2 \times 22 = 88 \div 2$ True   False

    C $10 \times 3 = 60 \div 4$ True   False

    D $5 \times 9 = 90 \div 2$ True   False

3.OA.3 DOK 1

2   The rectangle below is divided into square units. What is the area of the rectangle?

_____

3.MD.7A DOK 1

3   The perimeter of the figure below is 36 inches. Find the length of the missing measurement. Show your work, and give the correct answer.

10 inches

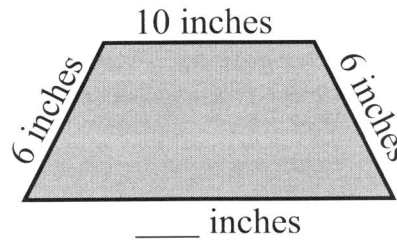

6 inches   6 inches

____ inches

_____

3.MD.8 DOK 2

Page 208

4   Look at the picture graph below. How many more students like raisin-oatmeal cookies for a snack than liked cheese for a snack?

Favorite Snacks of 3rd Graders					
Apples	🍎	🍎	🍎	🍎	
Cheese	🧀	🧀			
Raisin-Oatmeal Cookies	🍪	🍪	🍪	🍪	🍪

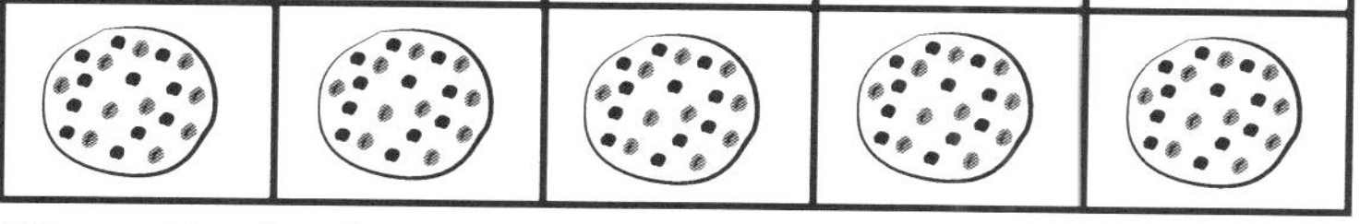

Key: Each picture = 5 students

A 3

B 7

C 15

D 10

3.MD.3 DOK 2

5   Luke wants to find out how much his dog weighs. What tool should Luke use to measure the weight of his dog? Write your answer on the line below.

3.MD.8 DOK 1

6   A grocery store is selling bananas by the bunch. Three of the bunches have 7 bananas, two of the bunches have 5 bananas, and four of the bunches have 6 bananas in them. How many bananas is the grocery store selling in all? Show your work, and write your answer on the line below.

3.OA.2 DOK 3

7 Circle all of the quadrilaterals below.

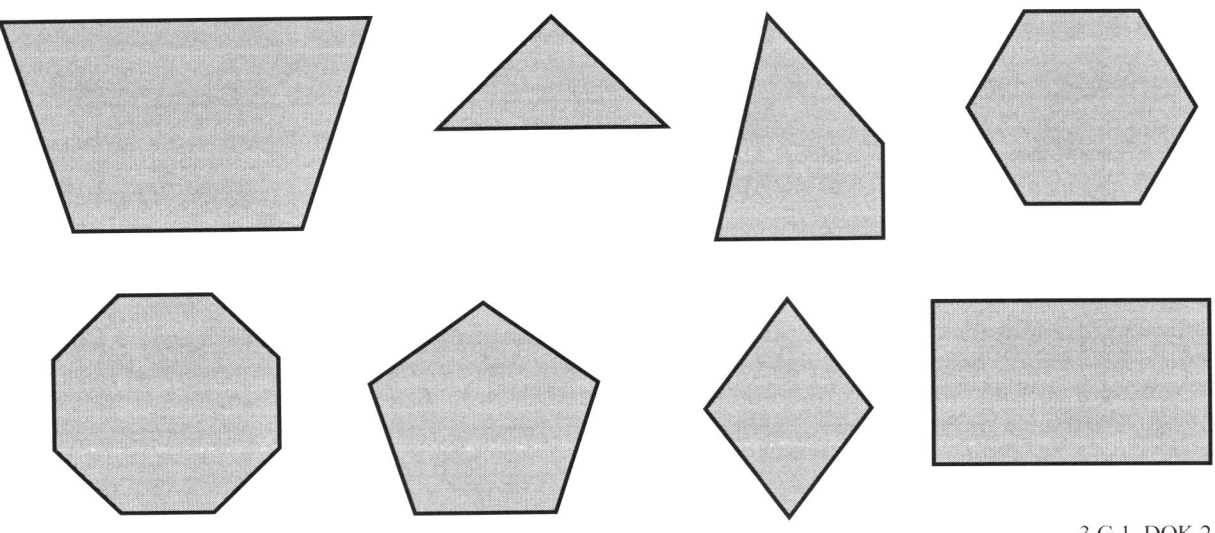

3.G.1 DOK 2

8 Which is equal to $(9 \times 5) + (9 \times 3)$? Choose the correct answer.

   A $(9 \times 5 \times 3)$

   B $9 + (5 \times 3)$

   C $(9 \times 9) + (5 \times 3)$

   D $9(5 + 3)$

3.OA.5 DOK 2

9 Round the number 5,505 to the nearest ten. Write your answer on the line below.

   _____

3.NBT.1 DOK 1

10 Which shape has fewer angles than a trapezoid? Choose the correct answer.

   A square

   B pentagon

   C rectangle

   D triangle

3.G.1 DOK 2

11 The chart below shows the number of pieces of fruit in a large box of fruit.

Fruit	Number of Pieces
Apples	10
Oranges	15
Pears	5
Plums	20

Answer each of the four questions below. Write your answer on the lines.

A What kind of fruit has 4 times as many pieces as the pears? _____

B What kind of fruit has half as many pieces as the plums? _____

C How many orange pieces and pear pieces are there in all? _____

D How many more plum pieces are there than apple pieces? _____

3.OA.7 DOK 3

12 Give the pattern rule for each row of numbers below. The first one is done for you. The answer for A is +2.

A 2, 4, 6, 8   <u>+2</u>

B 14, 11, 8, 5, 2   _____

C 5, 10, 15, 20, 25   _____

D 100, 98, 96, 94, 92 _____

E 84, 80, 76, 72, 68 _____

3.OA.9 DOK 2

Page 211

13 There are 24 brown sacks on the teacher's desk. Four of the brown sacks are empty. Each of the other brown sacks contains 6 jelly beans. How many jelly beans are there altogether? Show your work, and write your answer on the line below.

_____

3.OA.8 DOK 3

14 $82 \div N = 41$ and $N \times 16 = 32$. What is the value of N?

_____

3.OA.6 DOK 1

15 Which fraction is equal to 1? Circle the correct answer.

$$\frac{1}{5} \qquad \frac{5}{1} \qquad \frac{5}{5} \qquad \frac{4}{5}$$

3.NF.1 DOK 1

16 Ken was asked to subtract 9,722 minus 3,009. He lined up the two numbers over each other like this:

$$\begin{array}{r} 9{,}722 \\ -\,3{,}009 \\ \hline \end{array}$$

Which column should Ken subtract first? Mia said he should subtract the thousands column first because it is closest to the minus sign. Bob said he should start by subtracting the ones column first. Who is correct? Write the name on the line below.

_____

3.NBT.2 DOK 2

17 The drawing below gives the shape and measurements of a flower garden. How can the area of the flower garden be calculated? Find the area. Also, what assumption can be made about the drawing? Choose the correct answer.

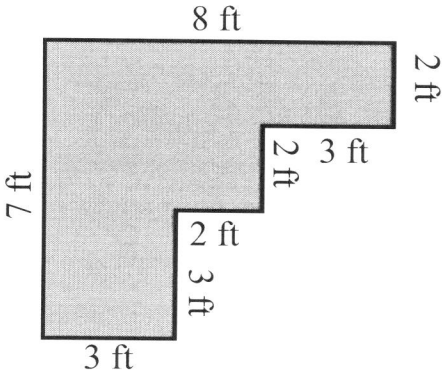

A You can calculate the area by dividing the drawing into rectangles and adding the area of each rectangle together. Area = 35 square feet. We can assume the lines are all straight.

B You can calculate the area by dividing the drawing into rectangles and adding the area of each rectangle together. Area = 35 square feet. We can assume the angles are all different.

C You can calculate the area by adding the numbers on the edges together. Area = 30 square feet. We can assume the lines are all straight.

D You can calculate the area by adding the numbers on the edges together. Area = 30 square feet. We can assume all the angles are obtuse.

3.MD.6 and 3.OA.1 DOK 2

18 Compare the two shapes below.

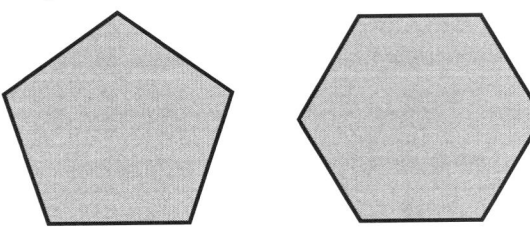

Choose the sentence below that correctly compares the two shapes.

A The hexagon has fewer sides than the pentagon.

B Both the pentagon and hexagon have an equal number of sides.

C The pentagon has fewer angles than the hexagon.

D Both the hexagon and the pentagon have the same number of angles.

3.G.1 DOK 2

19 Circle **all** of the fact families for the numbers 5, 9, and 45.

$45 - 9 = 34$ $5 \times 9 = 45$ $45 \div 5 = 9$

$9 \times 5 = 45$ $45 \div 9 = 5$ $45 + 5 = 50$

3.OA.4 DOK 1

20 Lisa's grandmother is making punch for her family. She uses the recipe as shown below and pours each ingredient into a large punch bowl.

Lemon Juice   **Orange Juice**   **Lemon-Lime Soda**

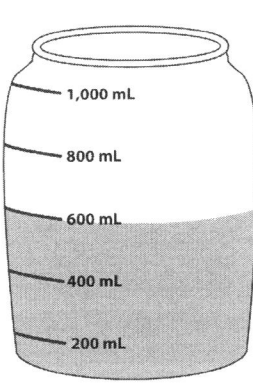
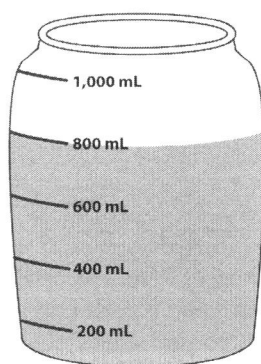

What is the total amount of liquid Lisa's grandmother used?

_____

3.OA.3 DOK 2

21 The bar graph below shows the favorite bugs of 3rd graders. Read the
    questions and write your answers on the lines.

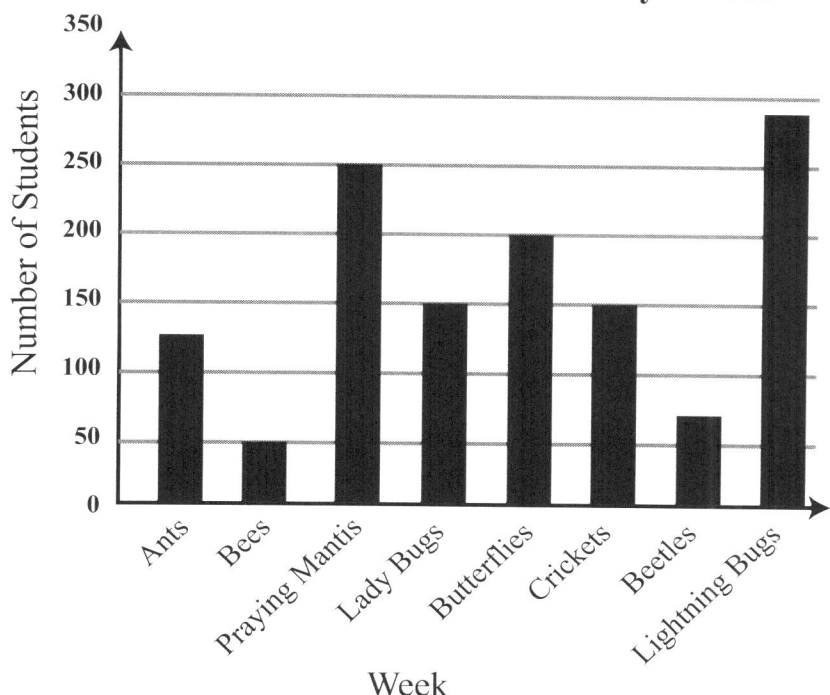

**Favorite Bugs of 3rd Graders
at I. Will Ketchum Elementary School**

A Which bug got more votes for favorite bug than any other bug?

_____

B How many more 3rd graders chose praying mantis as their favorite bug than
   bees?

_____

C How many 3rd graders chose butterflies?

_____

3.MD.3 DOK 3

22 The drawing below shows one wall in the Andrews' kitchen. What fraction of the wall is the window? Circle the correct answer.

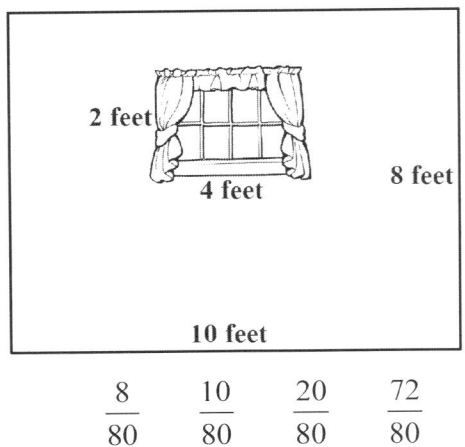

$\dfrac{8}{80}$ $\dfrac{10}{80}$ $\dfrac{20}{80}$ $\dfrac{72}{80}$

3.G.2 DOK 2

23 What is N in the problem $N - 4,009 = 3,782$? Choose the correct answer.

A $N = 227$

B $N = 7,791$

C $N = 8,001$

D $N = 7,781$

3.OA.8 and 3.NBT.2 DOK 1

24 A drawing of the upstairs hallway in the Jeffrey's home is shown below. The area of the hallway is 45 square feet. How long is the hallway? Show your work, and write your answer on the line below.

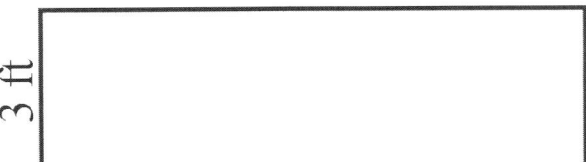

3.MD.7b DOK 2

25 Write two sentences below the rectangles: one sentence comparing the areas of the two rectangles and one sentence comparing the perimeter of the two rectangles.

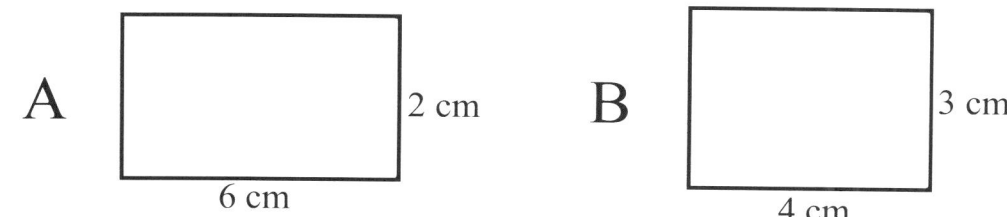

A   2 cm   B   3 cm

6 cm   4 cm

1) _____

2) _____

3.MD.8 DOK 4

26 Eric wants to know how many gummy candies are in 5 equal size packages. His friend, Bryan, says he should count how many gummy candies are in one package and then multiply by 5 to get the answer. His sister, Olivia, says he should count how many gummy candies are in one package and then add 5 to that number. Who is correct? Write the name on the line below.

_____

3.OA.5 DOK 2

27 Circle all the problems below that are solved correctly.

A $48 \div 6 = 8$

B $54 \div 7 = 6$

C $81 \div 9 = 8$

D $55 \div 5 = 11$

E $24 \div 6 = 4$

3.OA.7 DOK 1

28 Which of the following should be measured in grams, and which should be measured in kilograms? Circle the correct answer for each line.

      A a freight train filled with cows   **grams**   **kilograms**

      B one sock   **grams**   **kilograms**

      C a mini-van filled with people   **grams**   **kilograms**

      D a nest of sparrows   **grams**   **kilograms**

3.MD.2 DOK 1

29 Mark shaded $\frac{4}{6}$ of the hexagon below.

Circle all the shapes below that have an equal portion of area shaded as the one Mark shaded above.

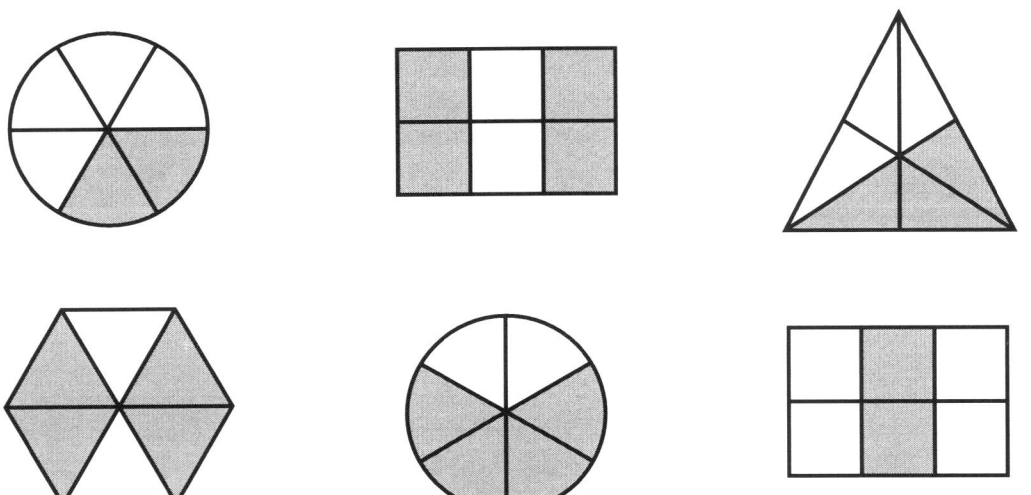

3.NF.1 DOK 1

30 Jill is trying to solve the problem (82 − 17) + 4. First, she added 82 + 17 = 99. Then, she added 4 so that 99 + 4 = 103. She did not get the correct answer. Where did she go wrong? Write a sentence telling where Jill went wrong, and give the correct answer on the lines below.

_____

_____

3.NBT.2 DOK 3

31 William is at the doctor's office for a check up. Before the doctor sees him, the nurse weighs him and finds his height. What tools does the nurse use for his weight and height? Write your answers on the lines below.

_____

_____

_____

3.MD.4 DOK 2

32 The line graph below shows the number of tickets sold by the Comical Circus from January through April. Write a sentence about the number of ticket sales from February to April on the lines below.

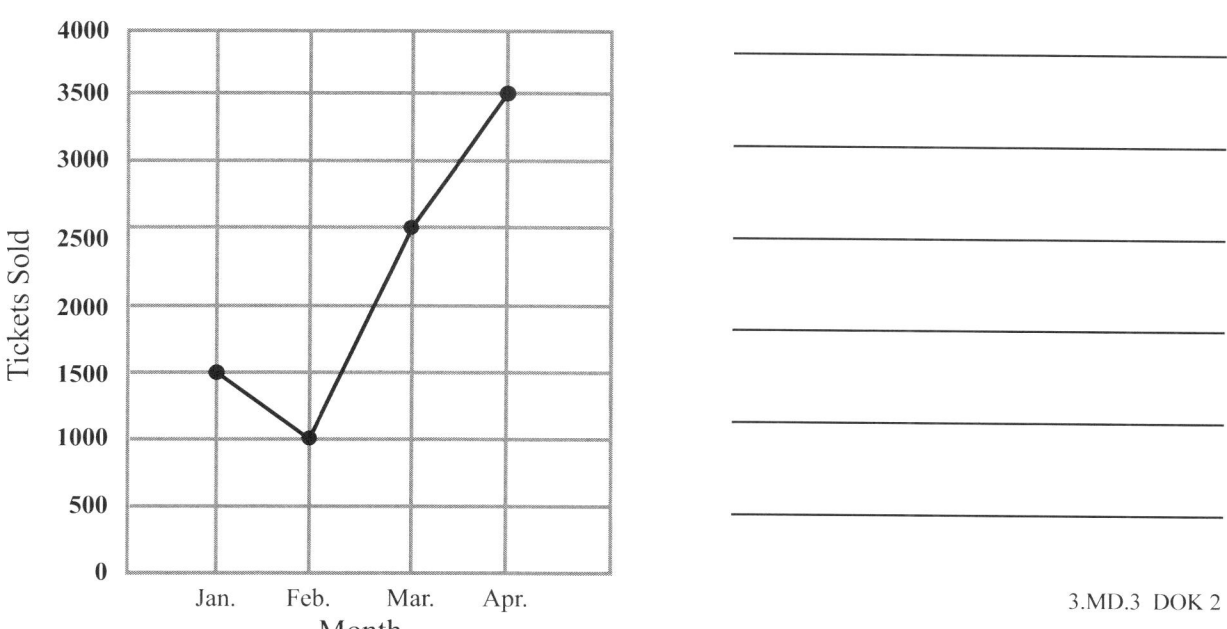

The Comical Circus Ticket Sales

_____

_____

_____

_____

_____

_____

3.MD.3 DOK 2

33 Suppose your teacher gave you a four-sided polygon made of paper. Without using a ruler or a protractor, verify that it is a rectangle. Write two things that define a rectangle on the lines below.

_____

_____

3.G.1 DOK 2

34 Find the perimeter of the polygon below. The length of each side of the triangles is 6 inches. Show your work, and write your answer on the lines below.

6 inches

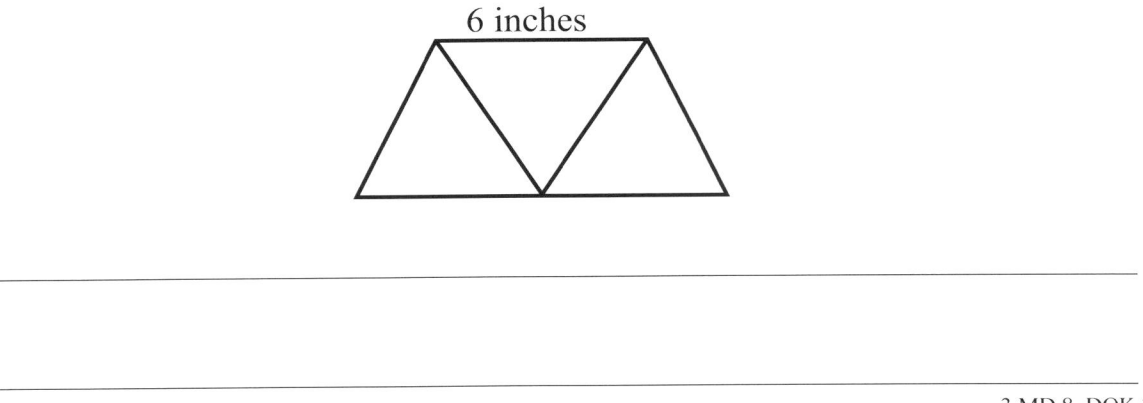

_____

_____

3.MD.8 DOK 3

Performance Task

You are taking part in a design competition. The goal of the competition is to design a tablet using specific rules. First, you must submit a drawing of the tablet with the following parts.

Draw your tablet on the following grid. Each square represents 1 square inch. Your drawing must contain labels that show you followed the rules

35 The tablet must be a rectangle with a perimeter of 36 inches and an area of 80 square inches.

3.MD.5 DOK 3

36 Your drawing of the tablet should be evenly divided, with $\frac{4}{5}$ of the parts shaded, indicating where the touch screen will be. The other $\frac{1}{5}$ of the parts will be where the control panel buttons are located.

3.G.2 DOK 3

37 Draw three control panel buttons that are quadrilaterals, but not rectangles.

3.G.1 DOK 1

38 A code is needed to unlock the screen of the tablet. Create this code by using a pattern. The pattern rule is to add 7. The first number is provided. Write the rest of the code in the spaces below.

3				

3.OA.9 DOK 2

39 The tablet will have 16 GB of internal memory. The following apps come preloaded onto the tablet:

App	Amount of Memory Used
Minesculptor	2 GB
Wakey-Wakey	1 GB
Movies Now	7 GB
Handspace	2 GB

Will there be enough memory left to load the app "SES Video Game Emulator" that uses 5 GB of memory? Show your work, and explain your reasoning.

3.OA.8 DOK 2

40 It is 2:15 p.m. and the design for the competition is due at 11:30 a.m. the next morning. Your friend says that you have 21 hours and 45 minutes to work on the project. Is he correct? If not, what is the correct amount of time you have to work on the project?

3.MD.1 DOK 2

Index